The Importance of Innovation Ecosystems in the New Space Industry:
A Swedish Perspective

Jackil Mann

Table of Contents

1 - Introduction

In this chapter, we present the importance and reasoning behind our choice of topic, while providing foundation and background for our research. Then, we will discuss the gaps in current literature, related to innovation ecosystems and the space industry, followed by the introduction of the research question as well as of its purpose. The closing part will elaborate on the intended contributions and delimitations of the book.

1.1 Choice of subject

Technology is taking over the world with strides unseen in human history and advancing at a rapid rate incomparable in time creating disruptive global changes in the space industry and the utilization of space data developing a paradigm shift in downstream and upstream markets that were once only under governmental control to dependence and cooperation among different entities and organizations through privatisation. This technological shift has opened doors to new developments, and application uses that demand space data leading to new markets and business opportunities that lead to increased global competition over technology and customers pushing governmental institutions worldwide to update their space legislations to retain competitiveness in their relevant space industries (Rymdstyrelsen, 2018, p. 4). In 2016, Nasa released its CubeSat Launch program. A CubeSat is a type of nano-satellites, measuring 10x10x11cm, making it far more accessible for CubeSat developers to conduct research, exploration, technology development (NASA Opens New CubeSat Opportunities for Low-Cost Space Exploration, 2015). Despite a large global investment of 360 billion USD (Global Space Industry Market and Technology Forecast to 2026, 2018) Sweden is only contributing 0.12 billion USD (Campanello, 2019) and is hence not a big global actor in the New Space industry. Lately there has been a increase in affordable satellite services which in turn has opened up the space industry to more actors, for example it is now possible to monitor factors which affect crop yield in the farming sector, in real estate satellite imagery can be used to identify sinkholes and in retail, satellite images can be used to analyze customer behaviour (Marr, 2017).

The space industry has always been an integral part of science, medicine, technology, and our everyday lives and its dominance lies with the creation and development of the latest technologies. A new example, the organization Planetary resources has since 2012 been attempting to harvest minerals from space and in 2015 the Space act law regarding privately mined space resources came to existence, essentially, making the finder of the mineral the owner, furthering the possibilities New Space brings (Capova, 2016, p. 308). Furthermore, technologies that are usually created within the space industry can be highly transferable to other industries furthering performance outputs through integrated technologies that have the potential to serve all stakeholders within its ecosystem. Space data is an ever-increasingly important source for knowledge-intensive organizations and companies producing advanced information products and services while also driving societal functions (Rymdstyrelsen, 2018, p. 5).

The space industry is highly interconnected with a country's technological capabilities, and its success is reliant on policies and regulations that nurture and promote growth with international cooperation ever-becoming prevalent to ensure sustainable practices, and safety. As space research is further developed to find answers to more questions, new disciplines are being added introducing new research funding agencies and cooperation with non-actors, and private actors due to the intensive requirements of resources (Rymdstyrelsen, 2018, p. 5). As the space race is ignited once again with the endless opportunities it presents with coordination between governmental and private actors through privatisation for New Space opportunities, opportunities that would not only serve for space expansion but rather serve earth-related projects and opportunities; their lies an importance to ensure an innovation-ecosystem is in place with policies and regulations that promote the overall performance, networks, and relationships that exist within its system; as an investment in space activity impacts society as a whole and is an investment for the planet.

1.2 Problem formulation

In this section we provide the direction in how we formulated our research problem through identifying how it relates to innovation ecosystems, the transition from old space and new space, and furthermore the adoption of new space functions in innovation ecosystems.

1.2.1 Innovation ecosystems

There is a debate regarding organizational ecosystems. Traditionally, the debate has been regarding activities inside and outside the boundaries of an organization and the outsourcing of production. Today much of the research has been focused on the attention of actors in networks who are involved in developing and commercializing innovations (Vasconcelos Gomes et al., 2018, p. 30). Ecosystems have according to (Valkokari, 2015, p. 17) emerged during the 1930s and were employed in social science, by viewing the economy as an entity where organizations and consumers are the living organisms. Moore (1997, p. 142-144) reintroduced the concept in the 1990s to make it relevant to discuss in management studies, who primarily wanted to utilize the organizational ecosystem to exploit self-organizing properties of natural ecosystems.

Management research on ecosystems can be separated into three different streams, the "business ecosystem" which puts focus on a organization and its environment, the "innovation ecosystem" which instead puts its attention to particular innovations or new value propositions and the network of actors which supports it and lastly "platform ecosystem" which regards how actors organize themselves in a technological platform. The innovation ecosystem definition does include and acknowledges that interdependence across actors exists and that they furthermore anchors the ecosystem to a specific "focal offer" or "focal value proposition" from the consumers point of view (Shipilov & Gower, 2020, p. 100). Adner (2006, p. 2) states that innovation ecosystems are "the collaborative arrangements through which firms combine their individual offerings into a coherent, customer-facing solution". Innovation ecosystems are enabled by information technology which has reduced the cost of coordination and made innovation ecosystems into a core element in growth strategies in many different industries.

There are three different "parts" of an innovation ecosystem: actors, artifacts and institutions and would according to Granstrand & Holgersson (2020, p. 3) be defined best the following: *"An innovation ecosystem is the evolving set of actors, activities, and artifacts, and the institutions and relations, including complementary and substitute relations, that are important for the innovative performance of an actors or a population of an actors"* as can be seen in figure 1 below. In the definition artifacts include products and services, tangible and intangible resources, technological and non-technological resources and other types of system inputs and outputs, including innovation. Meaning that an ecosystem can include an actor system with collaborative and competitive relations with or without a focal firm, an artifact system with complementary and substitute relations (Granstrand & Holgersson, 2020, p. 3-4).

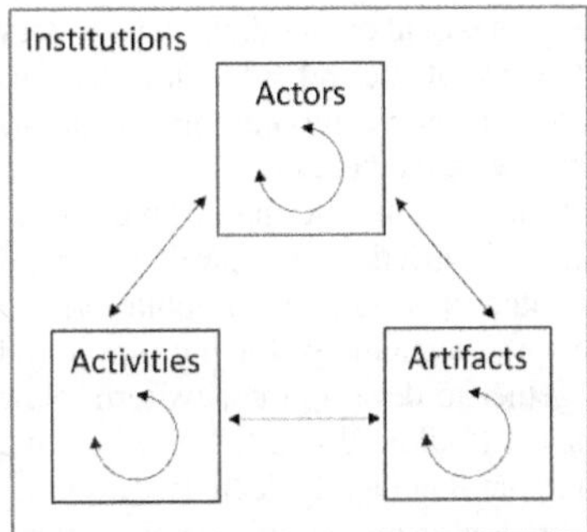

Figure 1: Illustration of the innovation ecosystem (Granstrand & Holgersson, 2020, p.7)

Adnan & Kapoor (2010, p. 306) argues in line with Granstrand & Holgersson (2020) without using the terminology, complex innovations often are not developed in a sole organization but rather also depend on innovations from other actors. For example, when Airbus launched the commercial jumbo jet A380 they faced substantial challenges related to designing and manufacturing the core airframe of the airplane. Beyond those internal challenges Airbus relied on suppliers for sub assembly and components. These suppliers are often themselves faced with innovation challenges to meet Airbus's demands regarding for example engine and navigation systems. Furthermore, Airbus is challenged with integrating the subcontracted parts onto their own core airframe. Even outside Airbus's supply chain, organizations are faced with challenges, airports need to innovate new infrastructure to facilitate new over-dimensioned aircrafts and simulator manufacturers need to innovate new simulation solutions where crews can be trained. Thus, Airbus is not alone in its innovation process but rather it involves a number of suppliers, buyers and complementors in the process (Vasconcelos Gomes et al., 2018, p. 307).

1.2.2 From "Old space" to "New space"

Since human space exploration started the ecosystem surrounding space has been under governmental control (Paikowsky, 2018, p. 84). The space activities were initially government financed and used as a tool to achieve national prestige and was the prominent way of funding space activities in the years 1945-1970 (Peeters, 2018, p 187). The ecosystem of "Old space" was created and shaped during the Cold War and even though the Cold War ended more than 25 years ago the ecosystem still lingers in the space industry. The primary stakeholders in the old space ecosystem were the superpowers and their close allies, motivated mainly by national considerations.

Space technology is dual use, meaning that it has both military and civilian use. In fact, much of the technology that was developed during the Cold War, the superpowers put strict restrictions on proliferation of know-how and launching technology as well as on other dual-use space technologies that were strategically valuable, such as high-resolution imagery, satellites, navigation systems, subsystems, and components. The superpowers reasoned that if other countries can use such technology it would render their own military advantage null. The result was a closed ecosystem due to strategic and prestigious reasons with state actors dominating and undermining commercial actors, thus commercial actors kept away (Paikowsky, 2018, p. 85). During the last few years there has however been a change and there has been a surge of private sector organizations entering this previously state-only undertaking, see figure 2 below. Strategic restrictions on proliferation of knowledge and technology were removed which created opportunities in public-private partnerships, whereby organizations could use the same technology for both civil and military applications. These changes have come together to develop the new term New Space". In the present time the two ecosystems coexist (Paikowsky, 2018, p. 84) while still the underlying principle still lingers. When public funding rapidly decreased after the moon landing, the major space companies had to directly or indirectly look for private space activities (Peeters, 2018, p. 187).

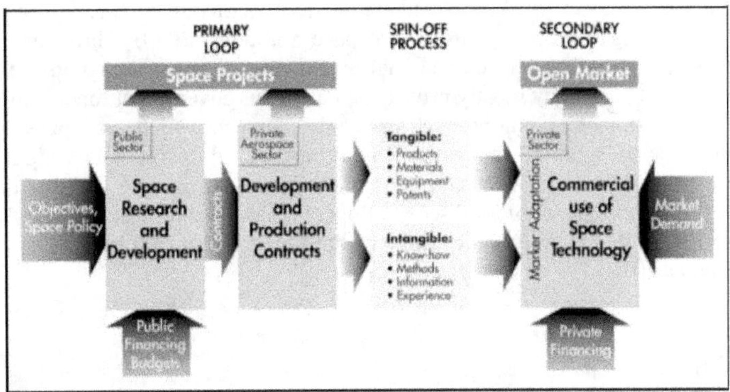

Figure 2: Transfer mechanism from public to private space activities. (Peeters, 2018, p. 188)

1.2.3 Innovation ecosystem in New space

A clear trend among the New Space ecosystem is the growth, development and use of small satellites. The number of satellites weighing up to 50 kg has increased drastically and now make up for a significant portion of satellites launched annually. The trend brings challenges about sustainability and safety regarding space environment, implications of crowded orbits, regulation issues, and their possible effects on the New Space ecosystem (Paikowsky, 2018, p. 86). To counteract the increased traffic in space, the United states, Australia, Great Britain and Canada have signed a memorandum regarding space situational awareness systems which provides data and warnings about potential collisions. Russia, and many other countries are developing new ways to focus their experience in improving their ability to identify objects in space. European space associations are working on a new type of telescope, currently more efficient at detecting debris. Japan is creating satellites which can return to earth, leaving no debris in space in the first place. China and France have passed new laws and regulations to prevent space from being overcrowded with debris. In the New Space ecosystem there will be greater regulation and standardization of demand from the private sector (Paikowsky, 2018, p. 87). Paikowsky (2018, p. 87) straight forwardly meaning that there will be a bottom-up demand from the commercial sector to pressure governments to coordinate space activity. Entrepreneurism and commercialism have entered space and with that there will be more energetic, creative, and dynamic systems in place rather than with "Old space". Governmental actors will together with nongovernmental actors have to learn to work together to face the problems and challenges that lie ahead. With a more open space there will be problems related to space traffic management, addressing the increased congestion in the electromagnetic spectrum, space debris, export controls and international cooperation (Paikowsky, 2018, p. 87).

New Space provides new possibilities with the use of satellite platforms for economically-driven activities with many fundamental uses that include satellite imagery and positioning, transportation, fleet management, gaming, virtual reality, apparel, textiles, personal safety, remote asset tracking, medical devices, agriculture, forestry, and environmental monitoring (Aaron Dinardi, Aarospace Solutions, 2020). Through the collaboration between new distinctly commercial organizations, aerospace companies, and associated ventures backed by mostly private capital versus government funds allow for a low-cost focus with incremental product/service development which also opens up for more risk-taking and spurs of innovation in contrast to old space that is highly dependent on cost-plus contracts with relatively no incentives to manage costs or do things differently outside of the requested scope of operations to avoid risks with failure and technical disasters that inhibits innovation within an ecosystem (Aaron Dinardi, Aarospace Solutions, 2020).

1.3 Research Gap

Based on our problem formulation and background in this section we highlight our research gap and how this research aims to add to existing academic literature.

The growing importance accredited to innovation as a mechanism of economic development has generated a large body of research to analyze the correlated relationships that exist between agents and their respective interactions. Prior research reflected heavily that innovation relies on distinct learning processes and the assimilation of knowledge creation, generation, and diffusion that are unevenly localized, thus the creation of hubs for innovative activity (Cooke, 2016, p. 2). Actors and parameters involved within such hubs for innovative activity and the relationships that exist between the different entities are comprised of governments and regulations, technological capabilities, supportive services, human capital, entrepreneurship, culture, market structures, and the local presence of research-oriented universities (Audretsch and Feldman, 1996; Fritsch, 2002; Stam, 2015).

According to past literature, such interactions and relationships are based on knowledge creation, generation, and diffusion but tend to overlook the connections that are presumed and its interconnectedness from a stakeholders perspective that drive knowledge sharing, networks, and implementation of innovative projects to succession. The stakeholders perspective theory contributes further to understanding innovation ecosystems, and provides foundational support to understanding how innovation ecosystems are created and formed with unified goals that fulfill each stakeholder's or entities ambitions on their journey to reaching a desired state or outcome, with recirculated benefits and outcomes throughout the innovation ecosystems network from gains that could not otherwise have developed retrospectively as a singular entity.

By furthering our understanding on innovation ecosystems from a stakeholders perspective, we can view innovation ecosystems from a point of perspective that all entities within the ecosystem are interdependent and that such entities within the ecosystem create value or should create value for each entities success or multi-laterally for all representatives within the ecosystem respectively (Freeman et. al, 2004, p. 365). Thus by understanding an innovation ecosystem from a stakeholders perspective we can attempt to grasp the motivation behind such cooperative work between different actors within the ecosystem, furthermore, potentially the driving factors that exhibit maximum efficiency and effectiveness within an ecosystem of interdependent actors/entities that drive innovation and co-creative solutions; which may be beneficial to furthering the space industry in the Kvarken region in Sweden.

1.4 Research question and purpose

Based on the identified research gaps outlined in the previous chapter, we developed the following research purpose and research question in this following section.

The main purpose of the book is to achieve insights in how the transitions from old space to New Space has changed the high-technology space industry and why companies who are ready for business opportunities in space related industries do not find them. More specifically how the Kvarken-region in Sweden and Finland is handling the new industry while identifying why space-ready companies are not in space related industries. According to Lantz (2020) this is an under researched area regarding the Swedish space industry. The New Space initiative is growing around the world but in Sweden we are seeing little growth and the aim is hence to understand why the growth is lacking and in extension how to stimulate growth. The book will attempt to fulfill its purpose by combining stakeholder, innovation ecosystem, opportunity forming, complementary , and institutional motivational academic theories.

The interest for the topic derives from the big global change regarding the space industry, and specifically the growth potential, total global expenditures was 360 billion USD in 2016 and is expected to grow 5.6% annually and estimated to be valued at 558 billion USD in 2026 (Global Space Industry Market and Technology Forecast to 2026, 2018); while Sweden contributes only approximately 0.12 billion USD (Campanello, 2019). Furthermore, the industry has seen a shift in what entities in the ecosystem drive research and implementation of space technology (Peeters, 2018, p 187). Lastly, there is as mentioned a disconnect between opportunities and seized opportunities regarding the space industry, as many companies who are eligible to conduct business in space believe "it is rocket science and not for us" (Lantz, 2020). These points make for a stable foundation as to why this research is interesting and why it has value for society.

Leading us to the following research question:

How do new entrants in the high-technology space industry navigate the transition from old space to new space?

1.5 Sample scope and delimitations

The main focus of this research paper is to understand how the Kvarken region has adopted new space sentiments in order to grow the space industry in Sweden. It is however difficult to completely remain in the Kvarken region as the industry is global. Markets and collaborations within the book tend to grasp outside our targeted region as a result. Moreover, the industry in Sweden is fairly small and even further insignificant in the Kvarken region, increasing the difficulty of optimizing empirical data as there are simply not that many organizations that fit our book design.

The theoretical framework has no direct correlation with Space thus our choice of theory relies on our assumptions with distinct regard to our empirical setting. Within this book an SME is considered as a start-up signifying their innovative and entrepreneurial nature compared to a traditional SME. Lastly, platform theory and stakeholder theory are regarded to address corporate perspectives, meaning the SME perspective is limited within those chapters.

2: Literature review

Within this chapter, we lay the foundations for our understanding of the relevant theories utilized in this research.

First, we will present the stakeholder's theory followed by an outline of the stakeholder identification process as many stakeholders maintain a position to decide the overall direction an SME will put into consideration.

Furthermore, we will identify fundamental factors of Platform theory which relates to stakeholder theory since platform partners will become stakeholders when a collaboration is established.

Therefore we will explore complementary innovation theory as it strongly relates to what type of collaboration is established for a firm.

Finally we explore motivational theories and institutional theories which help to understand why individuals seek the opportunities that they pursue.

2.1: Stakeholder theory

Stakeholders theory was developed in 1984 by researcher Edward Freeman whereby the researcher detailed the stakeholders theory for organizational management and business ethics in which it addresses morals and values within interconnected relationships between a business, and its customers, suppliers, employers, investors, communities, and others who have a stake in the organization. More importantly, how a firm creates value for all stakeholders, and not only shareholders (Freeman et. al, 2004, p. 365). As a basis with concrete foundations this theory has emerged throughout the years and has been utilized in an array of scholarly research and articles irrelevant of industries, or specializations to further evolving to capture economic, societal, and environmental value from an organizational perspective, or on an individual basis (Laplum et al., 2008, p. 1157).

The stakeholders perspective argues that companies and individuals play a vital role in the fabric of society and the world around us, and this role is dependent on value-creation with some ethical foundations and that success should be valued as a whole, rather than just on returns. This view that profit alone cannot be considered as the only measure of success intervenes that organizations and individuals need to center strategic viewpoints, relationships, networks, and value at the center of objectivity and strategy without overlooking each stake in isolation. Stakeholders are comprised of internal stakeholders, and external stakeholders (Laplum et al., 2008, p. 1157). Internal stakeholders for organizations according to Freeman's stakeholder theory which was later adapted by Deetz and served as sporadic inspiration for Mark-Herbert and Shantz (See figure 3) consists of management, employees, board of directors, and shareholders.

External stakeholders for organizations according to the stakeholder theory consists of customers, competitors, banks and investors, regulatory agencies, host communities, media, professional societies, unions, distributors, society at large, global ecological communities, suppliers, and trade associations.

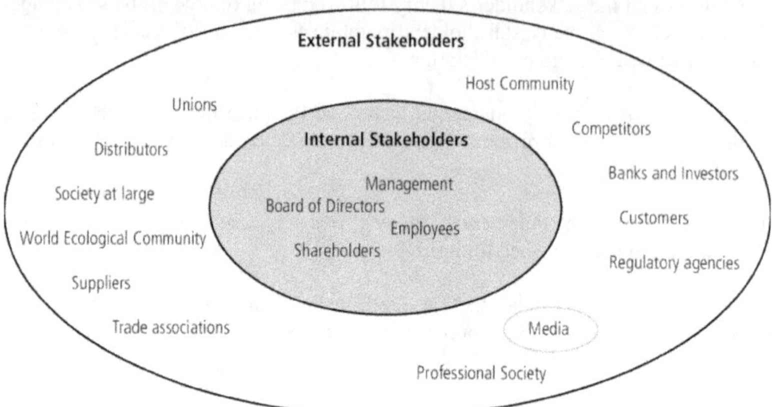

Figure 3: Model for external and internal stakeholders of a company (inspiration from Deetz 1995, 50-51) - (Mark-herbert & Schantz, 2007, p. 5).

The stakeholder identification process is very critical to an organization's understanding of their stakeholder environment, and the relative importance perceived and associated to each stakeholder. A firm understanding of their stakeholder environment allows for an organization to identify attributional models of really who matters and what really counts within their organizational environment (Mitchell et al., 1997, p.863). Stakeholder identification also helps furthering strategic scopes of influence and strategic operations that may or may not inhibit performance and executional outcomes. The two categories for a firm's strategy as identified by researcher Mario Minoja (2012, p. 71) are static solutions which are aimed at managing trade-offs and relieves certain sacrifices by sharing them among a majority of the stakeholders, or through dynamic solutions which consist of revising a firm's business model or even designing a new model that aims to connect strategy and ethics. The attributes presented in stakeholder identification are power, legitimacy, and urgency with each attribute's relative salience to stakeholder management and strategy (Mitchell et al., 1997, p. 869).

Power is a relationship among social actors whereby one actor could get or persuade another actor to execute an action they otherwise would not have considered solely through either coercive means (force/threat), utilitarian (material incentives), and or normative means (symbolic influences) (Mitchell et al. 1997, p. 865). The dimension of power in this context is referred to as the ability to mobilize significant resources to influence others which not only benefits large corporations that are usually projected with such abilities but also opens new forms of power for non-traditional stakeholders to utilize to create such influence (Jensen & Sandström, 2011 , p.474).

Globalization and the ever-prevalent interconnectedness of society has influenced stakeholder theory to recognize globalization's effects on organizations, and societies introducing the importance of new dimensions of power and responsibility (Jensen & Sandström, 2011, p.474). Power is also described to accumulate with many different complex networks and relationships that span and grow as an organization or as a stakeholder expands operations and put forth influence throughout the world. Through such global expansion, an organization will presumably expand its networks and relationship while also continually evolving and changing to meet environmental changes and new stakeholders within its capacity.

Therefore, a party to a relationship has power and can gain access to coercive, normative, or utilitarian means to impose will within the relationship and that the access to such means is variable, not a steady state and defines why power is transitory therefore gained and lost (Mitchell et al., 1997, p. 865). Furthermore, the creation of new power constructs and relational aspects in value creation models and how organizations attribute specific value, power, and responsibility among its stakeholders affects organizational strategy. Globalization brings with it a world of politics with many centres of authority and control with many transnational actors gaining more power, ranging from individuals on both spectrums of global corporations and non-governmental organizations as noted by Jensen & Sandström (2011, cited in Bauman 1999; Beck 2005).

The second attribute legitimacy consists of a generalized perception within some socially constructed systems of norms, values, beliefs, and definitions to either perceive or assume that an action of an entity is desirable, or socially accepted (Mitchell et al., 1997, p. 869). Such legitimacy can be based on individual, organizational, or societal interpretations thereby a system with multiple layers of analysis based on a social system (Mitchell et al., 1997, p.866). Legitimacy is described as being coupled with power to create authority, thus in terms of the legitimate use of power, further described if an entity as a legitimate stance in society or in a firm, unless coupled with power the entity may not be able to exercise one's will to achieve salience on the firm or society as noted by Mitchell et al. (1997, p.870, as cited in Weber, 1947). If achieving a legitimate stakeholders' claims creates impairment in a firm's long-term competitiveness, the survival of the firm itself is at risk therefore limiting its capacity to deliver value for stakeholders in the future (Minoja, 2012, p. 71). Therefore, an organization or firm is continually looking for static or dynamic solutions to manage legitimacy and power among its stakeholders to maintain effectiveness and efficiency among all entities.

The third attribute urgency is implemented to further understand the dynamic side of strategy, while legitimacy and power as independent variables are static in nature and go a long way in stakeholder identification and salience (Mitchell et al., 1997, p 867). Once a stakeholder needs or specific claim has been assessed and receives its relative importance a time dimension must be evaluated to meet such claims or needs in a specific time frame (Minoja, 2012, p. 72). This attribute of urgency consists of the degree to which stakeholders' claims require immediate attention and is based on time sensitivity in which managerial delay can be exercised and criticality the importance of the claim to the stakeholder (Mitchell et al., 1997, p. 869).

This idea of paying attention to various stakeholder relationships through time sensitivity is highly focused in issues of management and crisis management, and further an organization's capability in producing successful outcomes in managing such relationships (Mahon & Wartick, 2003, p. 25). Criticality when combined with time sensitivity assists in identifying a stakeholder claim or manager relationship as urgent or relatively unimportant (Mitchell et al., 1997, p. 867). Criticality is further described by Mitchell et al. (1997, p. 867) as four elements of observation which consist of ownership whether in terms of a stakeholder's possession of firm-specific assets or the assets tied to a firm that can not be used in the same efficiency and effectiveness without the relationship. The second element of sentiment in terms of relationships or stocks that are held by multiple generations of family members irrelevant of performance. The third element of expectation describes the connection between a stakeholder's anticipation that an organization will continue to provide it with great value and benefits. The fourth element is exposure which is the importance a stakeholder associates itself with that may or may not be at risk with the relationship with the organization or firm.

Further attributes to consider with stakeholder identification in identifying dynamism and salience with a firm as described by Mitchell et al. (1997, p. 868) are:

1. Stakeholder attributes are variable, not steady state.

2. Stakeholder attributes are socially constructed, not objective, reality.

3. Consciousness and wilful exercise may or may not be present.

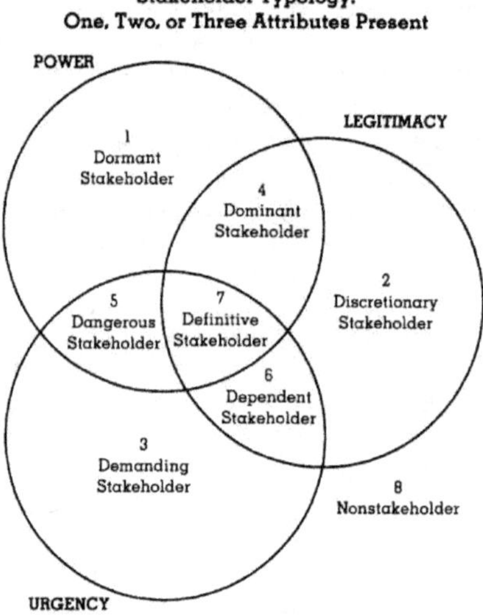

Figure 4: Stakeholder Typology (Mitchell et al. 1997, p. 872).

2.2 Industry platform theory

What managers and researchers refer to as platforms, can be found evident irrelevant of industries and come in many forms and shapes distinctly categorized among technology platforms, computing platforms, utility platforms, interaction networks, marketplaces, on-demand service platforms, content crowd-sourcing platforms, data harvesting platforms, and content distribution platforms (Michael and Sameer, 2020). Though such applications exist sporadically in different industries industry platform theory through its existence has developed touching on relationships and network effects. Furthermore its association with interactions is consistent of more users who adopt the platform. Therefore, positively reflecting among the entities within the network through the growing access to an expanding network within a platform creating potential for complementary innovations and incentives among the users and adopters of the platform (Gawer and Cusumano, 2014, p. 1).

Industry platforms and associated innovations can be embedded within other platforms, according to (Gawer and Cusumano, 2014, p. 2). An example of such a phenomenon is that " microprocessors embedded within personal computers or smartphones that access the Internet, on top of which search engines such as Google and social media networks such as Facebook exist, and on top of which applications operate, etc.)". Predominately in industry platforms, irrelevant of industrial background there lies the distinction between internal and external platforms (Gawer and Cusumano, 2014, p. 3). Internal platforms consisting of company or product platforms are recognized as a set of assets organized in a common structure that companies can use to create a stream of derivative products (Gawer and Cusumano, 2014).

External platforms consist of a group of products that are similar to the former innovation but provide a foundation for external actors, users, and outside firms to utilize in developing their own complementary products, services, or technologies within the business ecosystem (Gawer and Cusumano, 2014, p. 4).

2.2.1 Ecosystem theory

This concept of an ecosystem dates back to the 1930's and ecologists have focused on different definitions of the concept. Social science has for example long seen the global economy as an ecosystem where organizations and consumers are the living organisms. Moore fundamentally expressed in 1996, the reintroduction of this concept into management studies whereby the primary focus was directed through the actualization of the overall concept in self-organizinal properties in naturally occuring ecosystem that portray the business world as a means to gain advantages. Both man-made and natural ecosystems are always unique but share the fact that they both consist of a different set of actors and interactions and hence evolve in a dynamic manner and should not be viewed as a linear process.

The actors in each system have their own role to play which continuously shape the dynamics of the system. Ecosystems should not be viewed as platforms, but rather structures of relationships between the actors within and throughout the network (Valkokari, 2015, p. 17). Within management studies there are several overlapping structures consisting of business, innovation, and knowledge ecosystems. It is important to consider the boundaries of specific systems which can consist of a geographical scopes (local, regional or global), temporal scopes (historical or future), permeability scopes (open vs. closed); as the view significantly changes the dynamics.

Within each different system there are different businesses, knowledge exploitation opportunities, and innovational outcomes which are achieved by its actors through the combinition of different artefacts, skills, and ideas. Business ecosystems focus on how to create customer value, while knowledge ecosystems focus on generating new progressive knowledge, technologies, and innovation ecosystems that focus on how to integrate exploration (knowledge) and exploitation (business) ecosystems to foster innovation (Valkokari, 2015, p. 17-18).

	Business Ecosystems	Innovation Ecosystems	Knowledge Ecosystems
Baseline of Ecosystem	Resource exploitation for customer value	Co-creation of innovation	Knowledge exploration
Relationships and Connectivity	Global business relationships both competitive and co-operative	Geographically clustered actors, different levels of collaboration and openness	Decentralized and disturbed knowledge nodes, synergies through knowledge exchange
Actors and Roles	Suppliers, customers, and focal companies as a core, other actors more loosely involved	Innovation policymakers, local intermediators, innovation brokers, and funding organizations	Research institutes, innovators, and technology entrepreneurs serve as knowledge nodes
Logic of Action	A main actor that operates as a platform sharing resources, assets, and benefits or aggregates other actors together in the networked business operations	Geographically proximate actors interacting around hubs facilitated by intermediating actors	A large number of actors that are grouped around knowledge exchange or a central non-proprietary resource for the benefit of all actors

Figure 5: Business, Innovation, and Knowledge Ecosystems: How they differ and how to survive and thrive within them (Valkokari, 2015, p. 21).

This definition of innovation ecosystems is suggested by Valkokari (2015, p. 18) to be "regional clusters that focus on mechanism and policies fostering the creation of innovative startups around so-called regional hubs or clusters". That definition fits our book the best as we wish to explore how we can attract new companies to enter the Swedish space industry. Innovation ecosystems are focused on growth, interaction and innovation startups around knowledge hubs (Valkokari, 2015, p. 19). Clusters of innovation is a geographical area that facilitates the birth of innovative products and services and is supported by the ease of mobility of resources, including people, capital and information.

These clusters often see a large growth of companies, even in industries unrelated to each other. The clusters include for example startups; small, medium and large organizations; universities; research centers, investors; and service providers (Engel & del-Palacio, 2011, p. 32). Silicon valley is a successful example of an innovation hub for those primary reasons, conevenient flow of personnel, knowledge, and capital whereby it creates high expectations and integration of organizational development through the automatic introduction of more companies, overall benefiting the market while highlighting targeted synchronization to increase overall competitive progress (Claryssen et al., 2014, p. 1164).

2.2.2 Internal Platforms

Internal platforms are popularized as being in the context of reusable components and technologies that an organization, and its suppliers could cooperatively build and achieve to create a family of related products or to achieve incremental innovation to increase overall productivity and sales (Gawer and Cusumano, 2014, p. 3). The creation of a reusable foundation for product development and research requires management and planning, as described by Wheelwright and Clark (1992) various companies have developed product platforms to meet the needs of different customers simply by modifying, adding, or removing certain features that may increase productivity and complementary innovations for reusability.

According to Robertson and Ulrich (1998) they dive deeper to recognizing platforms as a collection of assets that include components, processes, knowledge, people, and relationships. From a marketing perspective, there is an initiative to move from portfolio thinking to platform thinking, to consider exploitation of the commonalities between a firm's offerings, markets and processes together to achieve leverage and business growth (Sawhney, 1998).

These researchers altogether have identified to an extent that there are several benefits to internal platforms that include savings in fixed costs, efficiency gains in product development through reusability and sustainability, the ability to produce a large number of derivative products with limited resources, and flexibility in product feature design (Gawer and Cusumano, 2014, p. 3). There is empirical evidence that shows that in practice companies efficiently utilized product platforms to increase product variety, control high production, and inventory costs reducing time to market (Gawer and Cusumano, 2014, p. 3). This research is mostly composed of durable goods with production processes in manufacturing such as the automotive, aircraft, equipment, manufacturing, and consumer electronics sectors (Gawer and Cusumano, 2014, p.3).

2.2.3 External Platforms

As described previously, external platforms consist of products, services, or technologies that are developed by more than one firm (Gawer and Cusumano, 2014, p. 4). There is similarity between internal and external platforms to the extent in providing a foundation for reusable products, components, and innovations, but it highly differentiates to how open or accessible the firm allows such reusability.

This degree of openness can vary on a number of dimensions, such as the level of access to information on interfaces to link the platform or utilize its capabilities, the types of governing rules on the utilization of such interfaces, components, or products and the cost of accessibility (patents or licensing fees) to such information, interfaces, platforms, components, and products (Gawer and Cusumano, 2014, p. 4). Despite the degree of openness various products and technologies can serve as industry platforms, this furthermore creates platform leaders, and complementors within a platform design (Gawer and Cusumano, 2014, p. 4).

A platform leader can be described as a firm that drives industry wide innovation for an evolving system of separately developed components that can be utilized by other users within the network, or platform (Gawer and Cusumano (2002, 2008). Gawer and Henderson (2007) described a product as "a platform when it is one component or subsystem of an evolving technological system, when it is strongly functionally interdependent with most of the other components of this system, and when end-user demand is for the overall system, so that there is no demand for components when they are isolated from the overall system".

The emergence to becoming platform leader and being capable of an industry-wide role and creating a stream of adopters and users to adopt the platform as their own there are two conditions reflected by researchers Gawer and Cusumano (2008) that must pertain to the platform. First, the platform must perform a function essential to the broader system, and secondly, solve a business problem for many businesses, and users in the industry. The researchers reflect further on the challenges of innovation dynamics whereby platform leaders and competitors will eventually face and navigate between competition and collaboration. Incentivization paradigms may occur as technology expands and develops, platform leaders may inherently want to enter developed complementary markets creating disincentives for complementors. Platform designs and strategic decisions for platform leaders have to be taken in a coherent manner, and fully discovered to understand the incentivization process and to ensure platform leadership.

---

1. Develop a vision of how a product, technology, or service could become an essential part of a larger business ecosystem
    a. Identify or design an element with platform potential (i.e., performing an essential function and easy for others to connect to)
    b. Identify third-party firms that could become complementors to your platform (think broadly, possibly in different markets and for different uses)
2. Build the right technical architecture and "connectors"
    a. Adopt a modular technical architecture, and in particular add connectors or interfaces so that other companies can build on the platform
    b. Share the intellectual property of these connectors to reduce complementors' costs to connect to the platform. This should incentivize and facilitate complementary innovation.
3. Build a coalition around the platform: Share the vision and rally complementors into cocreating a vibrant ecosystem together
    a. Articulate a set of mutually enhancing business models for different actors in the ecosystem
    b. Evangelize the merits and potentialities of the technical architecture
    c. Share risks with complementors
    d. Work (and keep working) on firm's legitimacy within the ecosystem. Gradually build up one's reputation as a neutral industry broker
    e. Work to develop a collective identity for ecosystem members
4. Evolve the platform while maintaining a central position and improving the ecosystem's vibrancy
    a. Keep innovating on the core, ensuring that it continues to provide an essential (and difficult to replace) function to the overall system, making it worthwhile for others to keep connecting to your platform
    b. Make long-term investments in industry coordination activities, whose fruits will create value for the whole ecosystem

---

Figure 6: Platform Leadership criteria and roadmap (Gawer et al. 2014, p. 429).

2.3 Complementary innovation theory

The role of innovation that is executed by external complementors is better described as complementary innovation, and it"s fit is utilized to assist firms and organizations in the construction of competitive advantages (Tsai, 2018, p. 1122). As mentioned previously, under industry platform theory, platforms and platform leaders depend on expanding the users and adopters of the platform; complementary theory dives further to understand the external platform environment whereby the degree of openness can be successfully orchestrated to increase innovation made by external complementors. Innovation management plays a key role in changing the strategic perception and direction of an organization through managing human capabilities, and the creation of networks with internal and external partners. Through the development and execution of adaptive and interactive organizational structures (Christensen and Overdorf, 2000). Innovation is defined as new and improved products, services, technologies or services that meet new requirements (Hidalgo and Jose, 2008). Internal and closed innovation which is better described as internal research and development is a great source for strategic assets, while managing external innovation assists in the creation of competitive advantages (Tsai, 2018, p. 1122).

Past literature on open innovation highlighted that purposive inflows and outflows of knowledge could support firms in improving internal innovation performance and the development of markets for external use of innovation (Tsai, 2018, p. 1122). Organizations could collaborate with external partners like universities, competing firms, and third-party developers to increase efficiency for innovation development (Tsai, 2018, p. 1123).

To fully comprehend open innovation, there are three key processes of open innovation which are categorized as outside-in or inbound innovation, inside-out, and coupled innovation. The outside-in approach focuses on how firms can make use of external knowledge to enrich their own knowledge base (West and Bogers, 2014). The inside-out or outbound approach focuses on how firms could transfer ideas and knowledge to the outside environment, while the coupled open innovation approach focuses on a combination of inbound and outbound approaches in the creation of open innovation (West and Bogers, 2014).

To further our understanding on complementary innovation, Teece, 1997 (p. 65) clarifies that there are two distinguishing factors; imitability and complementary assets; that will have a strong influence on determining which party or entity will profit from an innovation. Imitability refers to how easily competitors can copy a firm's technology or process that underlie the innovation (Teece, 1997, p. 63). There are barriers an organization could utilize to protect itself from imitation, swift action in intellectual property rights, complex internal routines, and or tacit knowledge (Teece, 1997, p. 60).

Complementary assets are equally important as barriers for protection and can include any activity that revolves around the core of innovation such as distribution channels, reputation, marketing capabilities, strategic alliances, customer relationships, licensing agreements, and among other elements that revolve around the core of innovation.

The following model explains further the relationships and predicts who will profit from innovation and also helps understanding which company will have higher incentives to invest in certain innovations.

Figure 7: Complementary assets/Imitability model (Teece, 1997, p. 67).

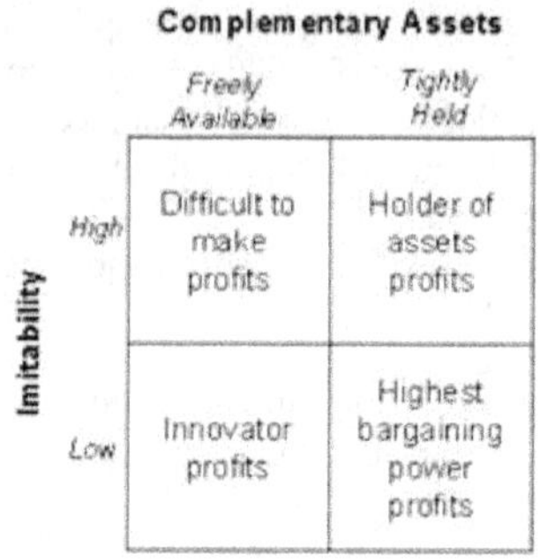

2.4 Opportunity forming

Institutional theory has been a popular theoretical foundation for exploring a wide range of topics such as institutional economics and political science to organization theory (DiMaggio & Powell, 1991, referred to in Bruton et al., 2010, p. 421).

Historically, the resource based theory of a firm has been one of the key theories as access to resources is vital for any new venture, in recent time it has been increasingly clear that issues regarding culture, legal environment, tradition and history in an industry and economic incentives cpuld all impact a certain industry and in turn new ventures (Bruton et al., 2010, p. 422).

Traditionally institutional theory is concerned in how organizations secure better positions within a market by conforming to rules and regulations within the institutional environment (Meyer & Rowan, 1991, p. 340). These rules and regulations derive from governmental agencies, laws, courts, professions, scripts and other social and cultural factors that implicate organizations in an objective way which actions are appropriate (Bruton et al., 2010, p. 422).

Meaning that institutional theory is concerned with more than efficiency but rather regulatory, social, and cultural influences which in turn will let organizations survive and thrive (DiMaggio & Powell, 1991, p. 130). The notion that institutional theory should include more than economic incentive has been suggested by Berrone et al (2010) who states that it is a feasible theory when exploring motivational patterns that go beyond the economical perspective. Shepherd & Patzelt (2011), motivationally argue that institutional theory has effects on sustainable contexts in both an economical and non-economical aspect.

In the following chapter, we aim to complete our literature review by including motivational factors into our phenomenon. A decision makers motivation comes from both internal (human motivation) and external (institutional setting) factors. The external motivational factors are strongly connected to stakeholder theory as stakeholders connected to an organization in question will likely in example; have opinions on how business is conducted, where resources are spent and which opportunities are relative to pursue. Distinctively, the theories presented in the following chapter have implications on platform theories.

2.5 Institutional motivation

The institutional theory is a complex theory due to the amount it adapts and encompasses theories from political science, economics, and sociology without having a clear focus (Scott, 2008). The theory indicates how organizations can accompany norms and rules of the institutional environment (Scott, 2008). The limiting rules and norms are created by governmental agencies and consists of laws, regulations, societal, cultural norms, and beliefs (DiMaggio & Powell, 1991). The evolution of the theory comes from institutional thinking, which was a way of perspectives and thinking that was adopted by Karl Marx, and Max Weber. They understood that an institution with rules and regulations will affect cognitive factors such as belief systems and knowledge sharing (Scott, 2008, p. 12-17).

In 1966, Berger & Luckmann made a direct statement that it was such thinking that has led to the development of institutional theory (Scott, 2008, p. 19-20). Robert K. Merton contributed to the theory in 1950, when he connected institutional studies to organizational studies (Scott, 2008, p. 19-20). Selznick built upon Merton's work and identified that planned or unplanned social actions are depending on the context (Scott, 2008, p. 20). Later in 1970, Simon & Silverman advanced the institutional theory further by stating that cognitive frames, rules and value have an impact on all individuals by creating meaning which is founded in institutional factors (Scott, 2008, p. 42).

Researchers in institutional theory have agreed that three different pillars exist in a continuum which drive the institutional theory: regulative pillar, normative pillar, and cultural-cognitive pillar (Scott, 2008, p. 50-51). The regulative pillar is most commonly perceived as being related to economics as it represents the rational thinker model of behaviour (Bruton et al., 2010, p. 422). Within the regulative pillar institutions guide and control actors by setting rules, laws, regulations; monitoring and creating sanctions (Scott, 2008, p. 52). This process can be achieved both by informal means, for example by shunning and formal means by the police or law (Scott, 2008, p. 54). The normative pillar is representative of the social dimensions contrary to the regulative one. The focus is on organizations and individuals behaviour which originates from professional, social and organizational interaction (Bruton et al., 2010, p. 422-423). The pillar shares the function with the regulative pillar in that it seeks to steer and guide how organizations should behave or act, the core is however "what is considered as proper" (Bruton et al., 2010, p. 423) rather by enforcing by law.

These social aspects can have both have motivational and discouraging effects on those seeking new solutions (Bruton et al., 2010, p. 423). The last pillar, cultural-cognitive factors encompass an individuals subjective rules rather than enforced laws or values like the previous pillars. The foundation of the pillar is in language, culture, already known and, taken-for-granted behaviour (Bruton et al., 2010, p. 423).

Scott (2008, p. 57) describes that the cultural-cognitive pillar is an external cultural framework which shapes internal explanations. In conclusion the institutional theory offers us a lens in which we can identify motivating factors for the opportunity forming possibilities that decision makers are seeking. Simultaneously, we could research how values and norms are perceived within the industry and to what extent it affects every entities decision-making process.

2.6 Human Motivation

 According to Parrish (2010, p. 510) there are two different streams of motivation when implying entrepreneurial decisions. The first relates to being motivated by profit while the other retrospectively sees sustainability, and profits are consumed as a side effect. Shane et al., (2003, p. 258; 263-269) divides and argues that nine different human motivational concepts impact the decision-making process for people who are pursuing entrepreneurial opportunities. Human motivational theories can explain why individuals seek certain entrepreneurial opportunities in which the correlation dictates the decision their companies take in the space industry. It is also tied to opportunity forming due to being concerned with how companies can make better decisions to secure a better competitive position in the market.

Table 1: Different human motivations.

Human motivational concepts	Explanation of the concepts
Need for achievement	Has a tendency to engage in undertakings which require a high degree of skill and effort while consuming moderate risk. Prone to wanting a high degree of individual responsibility.
Risk-Taking	Innovative undertakings can be risky and the decision maker can be motivated to pursue opportunities by the uncertain nature.
Ambiguity Tolerance	Decision makers making innovative decisions tend to have a tolerance for ambiguity and perceive it as attractive rather than threatening.
Locus of Control	The belief to which in extent a person's action affects the outcome. A decision maker with external locus of control believes that their actions do not affect the outcome while internal locus of control believe their actions do affect the outcome.
Self-efficacy	The belief that a decision maker has the ability to muster the personal resources, skills, and competencies to complete the task at hand.
Goal-setting	Having defined goals has a positive correlation to financial performance, growth, and innovation.
Independence	Pursuing innovative decisions that require self dependence due to the uncertainty revolving the outcome.
Drive	The length which the decision maker is ready to go in order to make an idea reality.
Egoistic Passion	The idea that motivation comes from a love for work.

2.7 Overview of the literature review

The first theory we present in our book regards human motivation as we see that as the chronologically first step and baseline for individuals or organizations to enter the space industry. It essentially translates in which ideas will be moved forward and which will be disregarded for other ideas. Secondly, we disclose institutional motivation as our standpoint is that the ideas and motivations generated in human motivation needs to align with institutional factors. The idea to enter the space sector needs to follow rules and regulations set in the market, as well as follow the social norms of "what is considered as proper". It is the cultural pillar of our book, meaning to explore why certain ideas can be adopted by the space industry and why other ideas have to go back and be re-evaluated to better fit within the industry.

These motivational factors moreover, have to be accepted by the stakeholders of the firm, which in turn can determine how and which business opportunities should be calculated and pursued overall impacting and morphing the motivational factors. Stakeholder theory is followed by industry platform theory, as many companies need to utilize different platforms in order to carry out their ideas and cannot act in a vacuum, this can for example be in the form of on-demand service platforms such as google earth, supplier platforms or, interaction networks. It is followed by ecosystem theory, due to the stakeholder decisions driving all ambitions to being apart of certain ecosystems and deciding on how they should integrate in such ecosystems.

It can be difficult to be in a "know-all" position for entrants in a technologically difficult sector like the space industry and in order to be fully adopted by the industry they may need to co-develop their ideas with others to fully develop a product or service which has demand on the market. Which brings us to internal platforms, exploring the idea on how companies can step out of the product/service portfolio and into the platform mentality to create a family of products/services, or whereby they achieve incremental innovation.

This strategy leads to positive effects such as increased product variety, the control of high production and reduction capabilities for timing to market. The following external platform theory is closely related as it concerns why a firm should be concerned with becoming a part of a broader system, which holds the end-user demand. This way there is no demand for the component or the sub-system itself but the demand itself stands with the evolving technical system which is being developed by a platform leader. Complementary theory then binds internal and external platform theory together by exploring how firms can utilize different streams of knowledge to achieve their platform status. Lastly we present opportunity forming theory as the last step of the process we believe impacts the reasoning why certain ideas are ultimately pursued on the market and adopted by the industry.

Figure 8: Summary of literature choices.

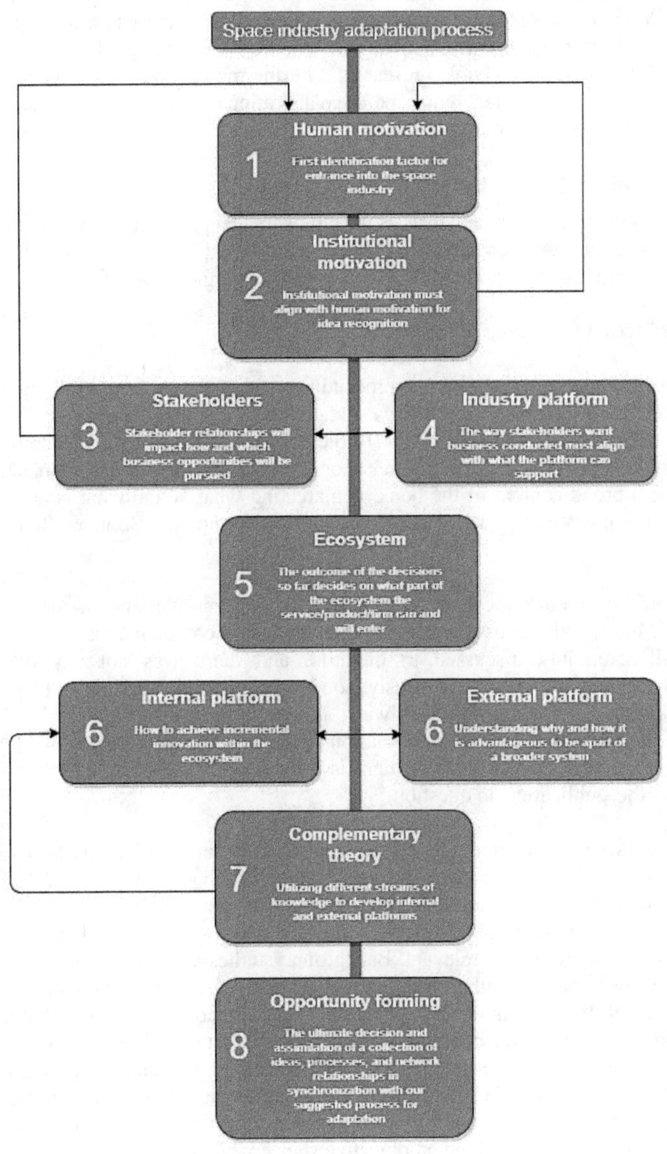

3 - Research methodology

Throughout this chapter, we present the methodological considerations on which we build our book as well as the methods for conducting the book. First, we will reflect on our choice of literature, and outline our philosophical stances. Afterwards, we will discuss the research approach and design whereby the strategy is explained with a unique focus on case book research. Furthermore, we introduce the data collection and analysis techniques before elaborating on the quality criteria and ethical considerations related to our research.

3.1 Philosophical perspectives

In this section we highlight and discuss the philosophical perspectives and stances chosen aligned with our research direction. Moreover, the explanation of our philosophical viewpoints and argumentation for the chosen methodological standpoints.

3.1.1 Reflection on choice of theory

A thorough literature review is a precondition for doing meaningful research. The researcher has to be aware of previous studies, what they mean and what strengths and weaknesses within the research exists (Boote & Beile, 2005, p. 3). The literature review is the foundation of any research and should be compiled of certain objectives such as setting a broad context of the book, demarcating what is and what is not within the scope of the investigation while also justifying those choices (Boote & Beile, 2005, p. 4).

Our choice of theory includes: stakeholder theory, motivational theory, ecosystem theory, institutional theory, platform theory, and complementary theory which were all commonly discussed in literature and there was not any difficulty in finding relevant sources in previously conducted research. Sachdeva (2008, p. 16) implies that it is important to always go to the original source, as it greatly reduces the likelihood of misinterpretation of the text and other errors. We have to the extent use original sources, even if there is a limit in databased information; we have access to the publication in question.

There are two paths a literature review can take, either it serves as background for an empirical book, or it is a stand alone piece of work (Xiao & Watson, 2019, p. 94). As the amount of research done in the areas is overwhelming to present although with the aim to serve as a background for our empirical studies, we chose between what we deemed either more or less relevant. Background studies are used to provide theoretical context, identify gaps, or otherwise justify decisions regarding research design (Xiao & Watson, 2019, p. 94). This course of action was pursued by presenting the traditional views on the theories and combining them with more up-to-date research in order to present the evolution of the theories used within each theoretical chapter. It goes well in line with what Xiao & Watson's (2019, p. 95) idea of a descriptive literature review, which aims to provide a state of the literature at the time of the review. It is worth noting that while we try to hold an objective stance while determining what theories are relevant and which are not, there are elements of subjectivity involved by natural definition as we chose certain ones over others.

Our literature search prior to the beginning of this research proved to be a good starting point as many authors, ideas, and academic research standing points were used. The sources utilized were found among Umeå Universities library database and Google scholar networks.

The main search phrases used and not limitied to throughout the research and literature search exploration was: **"Platform theory", "Stakeholder theory", "Motivational theory", "Complementary theory", "New Space", "Old Space", "Space industry Sweden", "Entrepreneurial motivation", "Institutional theory", "Ecosystem theory" and "Opportunity forming".**

3.1.2 Ontology

Long et al., (2000, p. 190) highlights that ontology "refers to assumptions held about the nature of a social reality". Key ontological questions are concerned if there is a social reality that exists independently of human conceptions and interpretation while also determines if it is concerned whether or not if there is a shared social reality or multiple ones (Ormston et al., 2013, p. 4). These two opposing views are called objectivism and constructionism; which are commonly taken philosophical stances within research (Bryman & Bell, 2015, p. 32). An objectivistic stance implies that social realities do indeed exist independently and create their own realities (Saunders et al., 2012, p. 130). The social reality is resistant to external phenomenons and the social entities are also unable to affect the phenomenon itself (Bryman & Bell, 2015, p.32).

Constructivism on the other side of the spectrum, is related in belief that each one of us is constructing a reality which is based on our own perception. Due to our view of the world being based on perceptions our constructions are imperfectly perfect (Sachdeva, 2008, p. 26). As we intend to book which internal and external factors contribute to entering the Swedish space ecosystem as a small or medium company we have adopted the constructivist view which assumes reality to be fluid, subjective, and constructed in its extension on the actors to create that reality (Bryman & Bell, 2011, p. 37; Saunders et al., 2012, p. 131-132). Further argument for taking a constructivism approach is that we wish to explore the interviewees opinions, which is grounded in previous experiences and their own understanding of the subject in matter.

While belief in their reality is to be through constructivism, it allows us to see how social actors are constantly shaping and reshaping their reality.

We argue that organizations within the New Space ecosystem are a social construct and not a naturally occurring phenomenon within the industry but rather a consequence of interaction between actors.

3.1.3 Epistemology

Long et al., (2000, p. 190) defines epistemology as "the basis of knowledge and in what manner knowledge can be transmitted to others". The two major epistemological stances are called positivism and interpretivism (Saunders et al., 2012, p. 134). Positivism can be traced back to René Descartes in 1637, whereby he underlined the importance of keeping an objective stance when searching for the truth. The key idea was that researchers should distance themselves from their subject so that the analytic process of the research will not be corrupted. The epistemological stance has since been refined in several iterations but the foundation still stands (Ormston, et al., 2008, p. 8). Positivism is most commonly used in natural science where the researcher plays a neutral role and does not impact the potential outcomes of research (Saunders et al., 2012, p. 134). The positivistic researcher sees reality as external and consisting of objective phenomenons, in which it consists of different realities independent on the researchers' beliefs or perspectives in the matter (Hudson & Ozanne, 1988, p. 511-512). That reality can be studied by observing the phenomenons without being influenced by predetermined theories and by developing methodology to test the hypotheses in the future (Bryman & Bell, 2015, p. 27-28).

With positivism takes the path of viewing reality as objective reality, the interpretivism has taken the opposite view and sees the social world as separate from the objective one and how it can be studied differently (Bryman & Bell, 2011, p. 16-17). Interpretivism was introduced by Immanuel Kant in 1781, who stated that researchers indeed can learn about the world by other means than direct observations. He meant that perception does not solely rely on sense but also human interpretation of what the senses tell us, meaning that 'understanding' arises from the reflection on what happened rather than just having a particular experience. Like the positivist stance, the interpretistic has been refined over the years but is founded in the philosophical standpoint that researchers should book people's 'lived experience' within historical and social contexts (Ormston et al., 2008, p. 11-12). By adopting an interpretivist approach to research when conducting social studies the researcher loses the richness and complexity of the phenomenons presented by Saunders et al ., 2012, p. 137).

The position taken for this book is interpretivism, due to the fact that data is collected through interviews and will hence consist of non-numerical data. Punch (2013, p. 17) states that interpretivism refers to understanding behaviour as people bring meaningful expressions to situations and thus the creation of an subjective sense of the world. Positivism on the other hand, researchers likely use theories to develop hypotheses and furthermore assume the reality to be observable and the data collected seen as law-like generalizations (Saunders et al., 2012, p. 137). However, as we consider people to have different views of realities and different ways of interaction and we wish to understand the perception of different stakeholders within the New Space ecosystem. Though the interpretivism path is more in line with the research that is conducted. The data will be open for interpretation from a social and historical context and not be law-like generalizations.
Meaning that we believe that interviewees will have their own subjective understanding of their reality and it could potentially be shaped by their unique interpretations of reality distinguished and known to them.

3.1.4 Axiology

Axiology refers to the role of values and ethics in research; Klenke et al., (2016, p. 17). Further described, is that dependence for a researchers value, is only on dependence of value and how it will affect the research is conducted and what the researchers achieves in the results of its book. Hogue (2011, p. 32) suggests that some researchers are striving to develop an understanding of how something works. For others, it is how people behave or to solve a problem. It could also be informing policy with the general aim to make the world a better place. Klenke et al., (2016, p. 17-18) describes that the traditional scientific approach is to conduct research that is unbiased and free of predetermined values.

That sentiment is not possible to achieve, all research is value laden and biased. Values tend to emerge from predispositions from disciplinary-related methodologies together with the researchers history and experience. The beliefs should hence be made clear to the respondents and the research stakeholders so it is apparent in which context the research is conducted and how it has been critically examined.

Since our book, conducts interviews we will be intertwined with the subjects in question and the level of bias could be seen as high as we determine what questions and follow up questions will be asked. Furthermore, we have chosen which companies and institutions we find relevant to include in the book, which is based on our value on what is relevant for the research being conducted.

3.1.5 Research approach

In any scientific undertaking, to ensure validity, replication, or understanding it is paramount to outline a research approach, meaning, to explain and justify the way that empirical material and theories co-create in the research (Bryman & Bell, 2012, p. 24).

In business research there are two main approaches: induction and deduction. Ghauri & Grønhaug (2010, p. 16) define deduction as "the logical process of deriving a conclusion from a known premise or something known as true"; and induction "the systematic process of establishing a general proposition on the basis of observation or particular facts". In quantitative research an inductive approach is most often depicted, which is a bit misleading as there is no "pure" inductive or "deductive" approach in research (Ormstom et al., 2013, p. 6). The authors argue that when research with an inductive approach generates and interprets data it is not possible to do with a blank mind.

The data generated, the type of questions asked and analytic categories will have the researchers influenced by assumptions deductively concluded from other work in the field. This also stands true for the deductive approach as when a hypothesis is tested it will originate from theory which has been reached by inductively reached by prior observations (Ormstom et al., 2013, p. 6).
This would also stand true for our research as our questions generated and the way the data is analysed has been generated from models and theories which have an inductive reasoning process. With that in mind, we have designed the book in an inductive manner and acknowledge that there will be inductive reasoning behind models and theories which have been used to generate interview questions and analytical tools.

3.1.6 Research design and strategy

For research with a qualitative approach, some of the most common methods used in the research design are case studies, ethnomethodology, ethnography, biographical method, interviewing, semiotic analysis, and observational techniques like observation by participation (Denzin & Lincoln, 1994, p. 13). In other words, there are several options to take into consideration. These are however only part of the first layer, defined as strategy. Saunders et al., (2009, p. 108) has created a model of what should be considered when designing research. This model shows different layers: strategy, choices, time horizon, and options that are available within each of the other layers. Table 2 below shows the structure that was chosen for this research.

Table 2: Summary philosophical positions.

Topic	Methodological choice	Comment
Ontology	Constructivism	The constructivist approach fits our research in the way that we want to explore the opinions grounded in previous experiences.
Epistemology	Interpretivism	We have chosen interpretivism as our epistemologicalstance. We acknowledge that certain objective realities might have shaped the perception of the interviewees.
Axiology	Value laden/Biased	We acknowledge our book will be biased and have values deriving from us. The questions and in interviews chosen the follow up questions we found relevant. We have also chosen which companies and institutions to include and exclude, depending on if we find them relevant or not.
Research approach	Inductive	We chose inductive reasoning for the book by acknowledging that our book will take an inductive research approach.

The purpose of this book is to find how more small and medium sized organizations can enter and thrive in a New Space ecosystem. To achieve that purpose we intend to conduct an exploratory case book, we will interview people who are in some way connected to the Kvarken region space industry. This book will be cross-sectional in nature, as the subject of the book is the New Space ecosystem in order to understand the ecosystem that exists. An exploratory case book underlines an empirical book of a phenomenon found within "reality" or "real life". This will be explained only through interviews, hence mono-method exploratory book.

3.2 Practical methodology

This section will focus on the procedures we have followed throughout our research in terms of data collection. We will lay out how we conducted our data collection while also putting into perspective our interview guide and the fundamental description for our chosen participants; thus detailing our interview guide and sampling technique.

3.2.1 Data collection

"Data collection is the very essence of any research" (Bryman & Bell, 2015, p. *12)*

Data collection is the process of collecting information through relevant sources that will assist the researcher to find answers to the targeted research problem. Such information and data will allow for hypotheses testing and the evaluation of outcomes. Data comes in many different forms and shapes but according to current literature it is categorized in two categories. Data equals primary data and secondary data. Primary data is data collected by the researchers depending on research design and approach. Our research takes a qualitative stance thus primarily the data is collected through semi-structured interviews, conducted observations, and an interview guide that reflects each recorded section with our participants. Our secondary data is collected from already existing data whether in terms of published materials on websites, newspapers, journals, and other documents to increase the overall effectiveness of the research (Saunders et al., 2012, pp. 304-307).

This secondary data collected within this book is retrieved from previously published academic articles, online sources, and previously conducted reports. The primary data is collected through semi-structured open interviews with key persons in small- and medium size organizations throughout Sweden; who have entered or are preemptively entering business opportunities related to the New Space industry. Primary data was also collected through observations on many key platforms that are already pushing for new space collective action in Sweden.

Our observations took place in Kiruna at the Space innovation forum in Sweden, and also through several workshops conducted in Sweden for new businesses and innovations to increase the new space ecosystem's initiatives. During interviews notes were taken, as Saunders et al., (2012, pp. 394-396) states it allows the interviewer to remain focused, making the interviewee feel their answers are important and let the researchers go back to a previous point during the interview if need arises.

The interviews all will be recorded if the respondent gives consent (See research ethics in 4.9). We hope that by having multiple sources from semi-structured interviews it will grant us a deeper understanding of the dynamics behind entering the new space industry for small and medium sized organizations.

3.2.2 Interview guide and Sampling technique

To ensure interviewees answer coherently and truthfully and to avoid irrelevant topics the book will be carefully explained to respondents before the interview begins. Furthermore, we will also clarify which timeframe that is of interest and whether the interviewee should refer to their own experience or the organization as a whole. The interview questions will derive from an operationalization of the chosen theoretical framework, to ensure we get the most out of the chosen theory. This will enable us to have a wide array of questions regarding opportunities in new space innovation ecosystems. The interviewee will also be notfied ahead of time of the interviewing process, so they are fully aware of the research purpose and what topics are of interest to us is for discussion.

Saunders et al., (2012, pp. 384-386) states it is important that interview questions reflect the subject being researched. However, the questions asked during the interview should not be too specific as it can prevent ideas from surfacing (Saunders et al., 2012, pp. 389-391). Which is why our supervisors have sent our topics prior to the interviews, as our supervisor has extensive experience regarding both research and our theoretical topics.

Tracy (2012, pp. 146-151) further highlights the importance of the interviewer being flexible with their line of questioning as the sequence of the predetermined questions can be changed depending on observations and answers during the interview. The result is open questions aimed at probing specific areas in which the respondent can define and describe situations to ensure significant results in relation to the research. After sending our proposed questions to our supervisor, the implementation moving forward is advised we scale back on the amount of questions due to the fact that it would be difficult to impossible to get through that amount of questions in the time frame of only 45 minutes.

The first iteration of the questions we formed for industry experts can be shown below in appendix 9.1. We also created a different set of questions for small and medium sized organizations which can be found in appendix 9.2. After revising the structure of our research questions we removed unnecessary parts which would enhance productivity. Moreovere, opens possibilities for us to find out for ourselves which we felt would serve us no purpose for the conclusion of the book. In contrast, we formed substitute questions which add further value to determine key factors which can be found in appendix 9.3.

The companies and experts that are carefully selected to fit our criteria as mentioned in our research method are as follows:

Company code	Position	Time in industry	Comments
SME1	Founder	21 years	Holds degrees in solid-state physics and a bachelors in economics. In 2010, established their first company offers space engineering capabilities. The second company aims to do environmental qualifications for start-ups space products.
SME2	Founder	21 years	Degree in industrial economics. Certified titles such as production manager and business developer until 2015, whereby they started thier own company. Specialized in antenna technology.
SME3	Founder	6 years	Certified degree in engineering, and physics. Based on the work accomplished in their masters book. Invented a unique technology that is now being offered to the space industry.
SME4	Founder	24 years	Master in navigational properties. Exclusive partnership/ownership of a company that explores innovative ways to use satellite imagery.
SME5	Marketing- and sales manager	2 years	No university degree. Works in the IT-sector and supports space companies with web-based development, system development, and cloud services for the space industry.
SME6	EVP	30 years	Degrees in electrical engineering. Works in a company specialized in electro- hydrodynamic pumps which are being used in satellites.
EXP1	Post doc researcher	5 years	PHD in physics. Involved in several projects and missions related to space research. Has seen several space companies emerge from the academic world.
EXP2	CEO	5 years	Masters in computer science and AI. Assisted IT startups for 20 years and in the 5 last years supported space companies.
EXP3	Technical director	17 years	Masters in electrical engineering, and a technical director at a Swedish institute providing funding for the space industry. Responsible for coordinating initiatives for the Swedish technology development in the space sector.

Table 3: Summary of interviewees.

3.2.3 Data analysis

Data is integral to any form of research, and the first instincts when receiving and collecting data is to find patterns, connections, and relationships. Once data has been collected the secondary phase would be to gain insights from the collected data. Data analysis is the phase in how researchers stream and decipher through mass amounts of data to create meaningful insights. Data can be analysed in several ways depending on the researcher's chosen path to conduct the research, mainly whether the research is quantitative, or qualitative.

For quantitative research, this analysis is mainly based on statistical evaluations, while qualitative data analysis is richer and in depth made up of words, observations, images, and symbols. The approach for this book in analysing the data collected was conducted with a thematic analysis, which is a method reliable in qualitative research for researchers who wish to use a low level of interpretation (Vaismoradi et al., 2013, p. 2).

Thematic analysis is a common method within qualitative research which helps the researchers to identify, analyse, and report patterns (themes) within data. It furthermore helps the reader to evaluate the research done, which is a difficult task if the reader does not understand the process of data analysis in the book. The method of thematic analysis raises the question of what qualifies as a theme. The answer is that the theme has to capture something important related to the research question. Braun & Clark (2006, p. 74) suggest that the researchers have to remain flexible when deciding what a theme within the book is. Rigid rules do not work and the researchers have to use their judgement in order to categorize the themes (Braun & Clark, p. 4-11).

Braun and Clark (2006, p. 87) suggests a 6 step process when conducting research according to thematic analysis. The process begins when the analyst begins to notice patterns of meaning and issues of potential within the data collected, meaning that it likely happens during the data collection. Since our research is inductive the use of literature during the early stages of analysis will be limited as that can narrow the analytic field of vision, leading to focus on certain aspects of the data and miss other crucial aspects (Braun & Clark, p. 74-76).

Table 4: The 6 phases of thematic analysis. (Braun & Clark, p. 87).

Phases	Steps	Braun and Clark suggested thematic analysis process	Our process
1	Familiarise yourself with your data	Verbal data has to be transcribed. Dotting down initial ideas for coding of data.	Transcription was not conducted manually. A program proved a solid foundation. The recorded interview was transcripted, and we marked interesting points, and corrected mistakes all mistakes. The transcriptions were completed after each interview.
2	Generating initial codes	Generated initial codes. Systematically worked through the entire dataset, giving each data equal attention.	Working with the transcriptions, we deciphered points that could end up being theïs thesis's empirical themes.
3	Searching for themes	The search for themes begins when all data has been coded. Sorting the codes generated from the data into themes. Organizing the codes into themes by using separate tools, such as mind maps or writing the codes in a separate paper.	In this step we organized all our data into data piles in a separate document. Here we started to get a clearer picture of how the themes would be presented.
4	Reviewing themes	Refinement of chosen themes. Two levels of reviewing, to check if the themes match the coded data extracts. Level one involves reviewing at the coded data extracts. Level two is to generate a thematic map.	Here we matched the coded data in our separate document to see if they match the chosen themes. Some themes did not have sufficient data and had to be collapsed into another theme or be discarded in the analysis.
5	Defining and naming themes	Defining and refining the chosen themes that will be presented. Identifying the "essence" of each theme. Identify each "story" that is related to the research question.	Further attempted to refine our themes to be relevant to the research question. Discarding irrelevant data from the transcription to not move outside of our chosen research question.
6	Producing the report	Making sure that the data presented is concise, coherent, logical and interesting. Convince the reader. Go beyond just providing data and make arguments in relation to research questions.	Choosing vivid examples from the data to compare with the chosen litterature. Seeing if the findings could be connected to the literature.

3.2.4 Interview process

The interview process consisted of 9 interviews. From these 9 interviews, 6 were conducted with SME's and 3 with industry experts from either academia, incubators or funding agencies closely tied to startups and SME's in the industry. The participants all have experience and expertise according to our criteria which is underlined in our methodology chapter. We chose to not disclose the identity of any participants as we wanted respondents to feel confident to disclose as much information as possible. The interviewees hence are coded, "SME" for organizations who are actively undertaking business opportunities related to the space industry and "EXP" for experts who are in academia, incubators, and or funding.

In table 3 below we have highlighted the basic information about the interviews.

Table 5: Summary of conducted interviews.

Code	Length	Medium	Recorded	Position
SME1	48 min	Zoom	Yes	Founder
SME2	42 min	Zoom	yes	Founder
SME3	57 min	Zoom	yes	Founder
SME4	54 min	Zoom	yes	Founder
SME5	66 min	Zoom	yes	CEO
SME6	59 min	Zoom	yes	Vice president
EXP1	47 min	Zoom	yes	Post doc researcher
EXP2	52 min	Zoom	yes	CEO
EXP3	42 min	Zoom	yes	technical director

The interviews were conducted from the 16th of April in 2020 until 8th of May in 2020. There were sometimes gaps of time between interviewees which allowed us to start working on the empirical findings and analysis in an iteration process so that it would not be an overwhelming amount of work as the book ran into the final phases. All of the interviews were done individually to prevent participants of the book from influencing each other. We would have prefered to conduct interviews in a personal manner as that is most in line with our methodological choices but due to a global pandemic and few to no organizations otherwise would not take personal meetings and a general reluctance to be in social spaces during the time of the book.

The interviews were aimed to be done in a 45 minute time frame, with an addition of 5 minutes per interview to clarify what our goal and purpose of the direction of the book while clarifying issues related to research ethics. Some respondents had less and some more to say when our questions were asked, so the actual time of the interviews would vary between 42 minutes to 66 minutes. All of the interviews were recorded which significantly increased our ability to focus on observing the responses as well as asking the right follow up questions rather than having to spend time taking notes. Recording the interviews moreover created convenience in transcribing the interviews which in turn helps with further analyzing the conclusion process

3.2.5 Summary practical methodology

Here we present a summary of the methodological choices that were made in our case book in an easy to overview table.

Table 6: Methodological choices.

Topic	Methodological choice	Comments
Data collection	Primary & Secondary data	The choice of using both primary and secondary data derives from the logic that while we are researching something that can be considered 'new' we believe the phenomenon can be analysed with existing theories.
Data analysis	Thematic analysis	The use of as little interpretation as possible but we recognize that research always has some level of interpretation in it.
Type of book	Exploratory Case-book	The need for our research to be connected to "real life" and hence decided to go with a case book on the swedish space industry.
Time horizon	Cross sectional	Given the time aspect of the research it is more fitting to conduct a cross-sectional book. A Longitudinal book is more consuming in both resources and time which we do not have.
Quality criteria	Non-generalizable but transferable	Generalizability and transferability are not mutually exclusive but can overlap. Transferability allows readers the option of applying results to outside contexts. Whereas generalizability is basically impossible because one person or a small group of people is not necessarily representative of the larger population.
Interview strategy	Semi structured open questions	The ability to explore the nuances of the interviewees opinions as well as to go deeper into certain answers, making semi-structured open questions more fitting for the research.

3.3 Quality criteria

In general, the quality criteria that is applied in qualitative research is different from the criteria utilized in quantitative research. This is due to the lack of strict rules when it comes to the data collection, analysis and interpretation. Therefore, the following categories, credibility, transferability, dependability, and confirmability, reflect on criteria that are of importance to qualitative research. (Guba & Lincoln, 1982, p. 426)

3.3.1 Reliability and Validity

Essential elements of any research book consist of quality criteria, a basis of analysis whereby the research has been conducted according to standards. Our qualitative research takes the approach of credibility, transferability, dependability, confirmability, and authenticity in terms of the instrumentation and results of the book (Riggs and Treharne, 2014, p. 58). Due to our research approach and design, these elements of quality criteria further enhance criticality towards the chosen direction of our research book.

3.3.2 Credibility

The credibility of our book is substantial (Riggs and Treharne, 2014, p.61); all participants included in our book were fully informed of the research book and are all currently activated within the Swedish space industry. Furthermore, we as researchers attended several conferences and events that are stimulating the current growth in the Swedish space industry. Through these conferences and events we were able to network with several small and medium sized organizations, experts, and university participants. Our participants were reached and contacted throughout our exploration of the current ecosystem that exists in Swedish space industry. Moreover, all participants were fully informed of our representation with Umeå university as masters students collaborating with the Kvarken space region project. All participants were informed of recorded interviews, and furthermore of transcription that has taken place, and sent for further analysis for the participants to ensure null discrepancies. All of the references in our work come from credible sources that were searched within Umeå universities library, and confirmed secondary sources that follow academic regulations and rules. All of our participants are registered businesses in Sweden, and have the appropriate credentials to conduct work and business activities within the space industry in Sweden.

3.3.3 Transferability

Due to the sensitive nature of our book correlated to fit the space industry in Sweden, our book is minimized when it comes to transferability (Riggs and Treharne, 2014, p. 62). Furthermore, our book can reflect on other capital intensive, and heavy infrastructure industries like the high-tech industry, or the automotive industry. But, due to the sensitive nature of the space industry and long returns on investment our book can be further transferable with further research on managerial decisions, organizational culture, and collaborative innovative practices in other regions.

Every country has its own regulations and rules when it comes to the space industry and its relative progress. Moreover, due to spill-over effects from cutting edge production and testing that persists in the space industry there always lies assimilation with cross functions of processes, materials, knowledge, and resources whereby similar industries can reflect and prosper.

3.3.4 Dependability

Similar outcomes in our research can be formulated by other researchers within Sweden (Riggs and Treharne, 2014, p. 61), with the accessibility to events and functions that are pushing the space industry in Sweden. The Swedish space industry is not relatively large like other geographic entities, thereby knowledge and access to participants active within the space industry further encompasses researchers and their networks. Our research is dependable due to our active participation in the conferences, workshops, and networks established and organized by confirmed entities within the Swedish Space industry. All transcriptions, notes, and recordings were further submitted to Umeå university in collaboration with the Kvarken space region project.

3.3.5 Confirmability

This research book as described throughout our research design and philosophical standpoints is an overview of experiences and knowledge received from our participants and their current experience within the space industry in Sweden (Riggs and Treharne, 2014, p.63). Furthermore, all references within our book are from recognized academic literature and reputable secondary resources that all follow regulatory factors in confirmable studies. Moreover, all data gathered within this book was well documented, analysed and shared with our participants' discretion.

We as researchers did not attempt to push our views and perspectives among the interviews taken place, and furthermore our participants have received full disclosure on their representative statements.

3.3.6 Research ethics

Ethics is one of many important considerations regarding research. In research it is how the subject is affected by what and how the researchers behave in relation to the data and rights of the subject. The ethical considerations range from everything from data collection to topic formulation and storing of data to research design. It is paramount the researchers act morally right in relation to the social norm (Saunders et al., 2009, pp. 183-184). There are a number of key phrases related to research ethics (Sachdeva, 2008, p. 31) *voluntary participation*, meaning that the subject should not be coerced into participating in the book. *Informed consent*, meaning that the subject has to be fully aware of the procedure and risks of the research and only then give their consent to take part of the book. *Confidentiality*, related to assuring the interviewees that the data collected will not be available to anyone who is not directly involved in the research.

This research targets SMEs and their thoughts about their opportunities regarding New Space; SMEs in general are more vulnerable and subjects were asked about strategic aspects of their organization. Full anonymity was offered and both name and organization was excluded from being mentioned. The agreement does not have any implications on the research itself but may affect the collection of data, as the trustworthiness goes down when information cannot be verified by a third party.

We saw it as a necessary precaution in order to be able to extract as much information from the interviewees as possible without jeopardizing their business or other undertakings, making the step reasonable. Saunders et al., (2009, p. 181) expresses that it is advantageous to offer several forms of interview methods in order for the respondents to be more likely to take part of the book and be granted some control over the situation. We only offered interviews through video conference software Zoom, and telephone interviews as the research was taking place at the same time as the peak of the COVID-19 pandemic. Essentially all organizations were reluctant to have face-to-face meetings due to health related concerns.

Saunders et al., (2009, p. 181) states it is important researchers are clear and specific regarding requirements of the book and what the aim of the book is. This is why when subjects were contacted we were clear about being students from Umeå University researching their organizational path from being a company outside the New Space industry to entering the New Space ecosystem. The interviews were also being recorded in order to have as complete data as possible, if the respondents gave permission to do so. The respondents were found on various lists of regional suppliers for the space industry, participants in space related conferences, lists of space companies who had been incubated and by references from previously interviewed respondents. All of the interviewees contacted were contacted directly by us and no third party involvement was used.

4 Empirical Findings

The following chapters contain a presentation of data gathered in our interviews with companies and individuals who are connected to the Swedish space industry. The first subchapter regards contextual information which is highly important for the reader to understand in order to process the rest of the data. The book follows a thematic analysis by Braun & Clerk (2006) and the data is hence presented in different themes gathered from the interviews with foundation in our literature review with the goal to help answer our research question.

4.1 **Contextual information**

Table 7 below, explains the different streams of interactions accumulated by our interviewees. Its importance lies among the distinction it creates, fundamentally two very different markets within the same sector. According to the interviews there are several different types of markets: downstream, data exploitation, new space, old space, upstream, legacy space, traditional space, and or infrastructure; which hence will be named downstream and upstream for the sake of consistency.

Downstream relates to the utilization of data coming from applications which are already in space, such as GPS, satellite imagery, weather data, satellite broadband, etcetera. Upstream or infrastructure relates to everything meant to leave the atmosphere, generally to serve a downstream purpose, further described as products that are meant to go to the space station, a rocket, or spacesuits.

We believe it makes it easier for the reader to have a distinctive and clear line between both entities for the formation of the two different branches in how they conduct business differs between them.

Table 7: Summary of interviewee organisations.

Company code	Branch	Area of operations	Comments
SME1	Upstream	Qualified products going to space.	Supported functions for upstream companies rather than an actual upstream company.
SME2	Downstream	Provides broadband services.	Strictly downstream.
SME3	Upstream	Developed robots meant for space missions.	Established themselves on the upstream market segment.
SME4	Downstream	Developed services by using satellite imagery	Strictly downstream.
SME5	Downstream	Developed IT-services for companies within the space sector	More of a supporting function for downstream companies than an actual downstream company.
SME6	Upstream	Produced and developed electro hydrodynamic pumps	Heavily established on the upstream market.
EXP1	Both	Conducted space research	Conducted research with relation to both upstream and downstream space activities
EXP2	Both	Incubated space companies	Does not directly act in a branch but engages with companies in both upstream and downstream space industries.
EXP3	Mainly upstream	Funding agency	Personally involved mainly in upstream space funding but the organization is concerned with both.

As can be seen in the previous table the SMEs have a split of 3/3 between being upstream and downstream companies which we believe to give us a better picture of the industry as a whole and how they relate to each other. Furthermore our three experts have different experience and knowledge about both of the branches which we believe to further strengthen the data collected as the interviews can be backed up with solid reasoning from experts with solid experience.

4.2 The Swedish space sector related to SMEs

It is a generally conveyed but not an unanimous message by our interviewees that the Swedish efforts in the space sector and in particular regarding SMEs are lagging behind other more effective strategies utilized by other countries. There has however been advancements within the space industry related to SMEs in recent years. The government appointed an overseeing funding agency that has gone from allocating 99% of their budget to the big four companies to form infrastructure and necessary requirements to initiate new phases to allocating 86% of their budget. Which therefore, significantly increases the allocation of resources to SMEs who wish to pursue space related undertakings (EXP3).

There are furthermore new rules that dictate that a certain percentage of the system integrator has to contain efforts from SMEs while also including rules about certain amounts of technology development has to come from SMEs (EXP3). The government also has taken a bigger interest in the space sector which could spark new and interesting changes for the ecosystem in the time to come. While this is an improvement on paper for smaller companies who want to engage in the ecosystem it does not mean much in practice according to several companies interviewed.

SME6 & SME1 both state that even though more of the funding budget is allocated for smaller companies it more often than not ends up with the larger corporations constranting further the dictation on what the smaller members of the ecosystem should provide for them in terms of services and products while indirectly gaining access to the supposedly allocated funds to the SMEs. SME3 does however agree with the structure of big corporations receiving the majority of the funds and stating that he believes that it is usually how innovation is done within the space industry.

"They have already established their contacts, so in depth and discussed the project Between their company and the space partners. So in depth that the space partner has already chosen them [...] the small company will never get this opportunity because they cannot prove that they can deliver on such a big promise. So they will get one meeting and then nothing more." (SME3)

The statement aligns well with what SME2 has observed and states:

"To enter the upstream sector is very cumbersome as old relationships based on decades of trust linger on within the industry hindering the ability for newcomers to enter with their business offerings, even though they may show promise." (SME2)

EXP2 confirms with an aligning view:

"The culture has been such that the Swedish national space agency has been working close to the Swedish space corporation. So there have been lots of large actors, but they have all been doing business among themselves. It has been really hard for a small startup to come in and supply stuff to these big vendors. How can a startup with three people start selling to Airbus?" (EXP2)

He continues to explain

"The buyers are the European space agency and large government organization, they like their procurements. They say that you need to supply me with a five year contract with validation that you have been around for 20 years and you got to be around for another 20 years [...] It has been rigged towards large organization buys from other large organizations. It has been tough for smaller companies to enter this. IT, telecom and the whole mobile industry has been totally the opposite with the internet just opening up in the last 20 years where it's even easier for small vendors to come in." (EXP2)

EXP3 has identified the same concept and states:

"You can only deliver hardware to a satellite if you have flown before. And if you have not flown, you could not have delivered to speak. It is a story of at least 10 to 15 years before you get accepted and that requires some stamina from the company in order to be able to endure this process". (EXP3)

This undoubtedly has caused severe problems for newly established companies on the upstream side of the space industry. They have to prove themselves for years to be able to be considered a serious actor by the established industry, suggesting it is near impossible to start off as an upstream company without a broader product portfolio targeting other industries or having serious financial backing, also coming back to why private equity is prioritizing other areas of investment.

According to SME1, there is an ongoing improvement to historical structures and the current trends shows promise, but just not as much as it could compared to other countries with enhanced processes. There is currently a review of the Swedish space legislation in progress on a government level, so things will formally adapt and change (EXP3). One of the experts hopes that the space industry will be identified as a high growth sector, similar to what the Medtech and the AI industry in Sweden has seen. Moreover in turn reproduce the success story that Sweden has had specifically to those industries.

EXP3 admits that Sweden has not been very efficient in data exploitation or downstream markets and specialized areas which is generally where most of the SMEs act but that could very well be changed coming in relation to the review of the legislation and its progression. Furthermore, the expert in the funding organization said that they could easily double the current budget and still distribute the money efficiently to different actors that have interesting and innovative ideas.

He also states that there is a human resource problem within their organization and that they simply do not have the labor force to achieve the results they intended. Sweden is also relatively low on private investment and capital within the sector due to lack of competition and opportunities which also contribute further on to why Sweden is lagging behind on the global scene.

"There is also a general lack of private funding in the industry compared to other countries, many investors see the space industry as uncertain and with high risk associated." (EXP2)

According to EXP2, the circumstances seem to be changing as we speak and more private equity is reaching the smaller companies every year, which according to him is a positive development.

There are also general resource restraints associated with the space industry for smaller companies due to the industry being global which the ability to compete highly refers to companies traveling to all parts of the global space industry and to fairs and conferences all over the world to keep up with the network building and knowledge gathering, but unfortunately an SME's budget it can be a difficult hurdle to overcome and they fall behind.

SME1 summarizes it accordingly:

"The market is still at large institutions and there are lots of costs associated with being in the industry. The market is global so you have to travel much so just keeping your travel budget is costly. If you have a really good idea, technical innovation, you can sell that to the space board and get funding through ESA and some other international programs and that is the way that you can succeed today" (SME1).

The possibility of trying to penetrate the upstream market seems close to impossible for SME1 unless there is significant change in the current market environment. Explaining why the ones that succeed on average are having a downstream focus.

There is opportunity to improve the current infrastructure to improve the overall productivity of the space ecosystem, and that such developments take time to assess, develop, and implement. While also the current "monopoly" that exists creates deficiencies in progress as most of the funding is distinguished to the big players, the larger organizations that have been around since the emergence of the space industry which therefore limits opportunities for new players and small businesses to fully integrate under the space platform.

SME6 believes that even though new opportunities have been created to lower the barrier of entry for small and medium organizations that the most successful outcomes in the downstream branch are the traditional players.

This success is further described as being dependent on industrial and manufacturing capabilities that traditional large organizations have access to and that the small and medium sized organizations simply do not have the resources due to capital-intensive requirements.

"Because if you are talking about the large amount of money in the downstream sector, you talk about large telecommunication satellite systems with several thousand satellites. There you need industrial capability [...] I would say a lot of those very small satellites, cubeSATs and so on, have not really succeeded business-wise." (SME6)

4.3 Collaboration

The majority of the interviews conducted on our participating SMEs identify that there is in one way or another collaborations and/or partnerships with other organizations. When a collaboration happens it does not only give a boost to an SME but furthermore reflecting further on turnover moreover it also increases the innovation pace in the whole ecosystem while also allows for organization to be more relevant global challengers taking new market shares as the space sector is generally speaking a global one. The reasons mentioned behind the collaborations from an SME point of view vary between the firms further reasoning includes but is not limited to knowledge gathering, synergy effects, R&D, cost reduction, and strategic competitive advantages.

SME5 wants to reach more customers and has a very open view on collaborating with different organizations, even those who otherwise would be seen as competitors.

They have a societal call which sometimes trumps the economic incentive.

SME5 summarizes by stating:

"We have a very pragmatic view on collaborating. We don't have any problem at all collaborating with a company that normally would be our competitor. We can combine our resources doing a project for a customer X, then that's perfectly fine and we are very open and straightforward." (SME5)

SME4 likewise collaborates with different research institutes to create proposals and to create new services that otherwise would be missed business opportunities. They further explain that collaboration with researchers is paramount in order to identify what is currently possible to develop and how the development is progressing in different fields.

They have a wish to develop and explore more than they do, but often small organizations often run into the problem of insufficient funds but they also identify the possibility to collaborate with other firms to do cost-sharing when developing new services.

"We had an idea together with x, y and some other companies regarding monitoring wild boar damages on farms [...] We had to find new ideas [...] It is very fun to add and try new things and invent and find new opportunities. But we also have to afford it. It could be interesting being a part of a large company which has full freedom to do whatever they want, see the interest in developing something new without the money restraint while also adding value to customers". (SME4)

Furthermore, SME5 collaborates with different municipalities due to new laws and regulations regarding environmental and climate change which they follow to in turn create new opportunities to make tailor-made solutions that fill those gaps in the market.

"There are new rules, regulations for the municipalities they have to follow and they have to take into account these questions in their planning processes. We can help them with that..". (SME5)

Hinting that it is important to develop your services for a specific customer in mind as they would miss out on the business opportunity if they were to offer a generic service that did not completely fit their demand. That is a recurring story in our interviews, interviewees several times mention that the offer they create must fit the core system of their consumers.

SME6 like SME5 develops a product with a specific system in mind further stated as:

"I would say for space products it is more or less totally in-house development. But of course we try to do it together with the customer because the product has to be designed into the system level design". (SME6)

They are more light in terms of collaboration with other space undertakings simply put the opposite of SME5, but they do however acknowledge the importance of fitting into the customers product and that their product offering cannot be a generic offering.

The founder of SME3 states that a collaboration with the robotics lab has been crucial for their existence. They showed a lot of interest in the development of a lightweight robotics arm which they then helped to co-develop with a larger organisation drastically reducing the time to market and cost of R&D.

"The robotics lab provided funding and they are really interested in this technology and see a great future for it. So they have helped with personnel and lent us space for a long time and let us develop in their lab". (SME3)

SME1 has identified possible positive outcomes from collaborations with geographically different firms and are looking for collaborations with the aim to share resources and gain access to new networks within the ecosystem.

"We have a lead out to collaborate with companies in Kiruna since we are located in Stockholm and you really need someone on the ground to get some activity up there. We are looking for someone who has already established themselves in Kiruna and help us to establish ourselves too and also share resources. Both to gain access to a new network. So it is win-win". (SME1)

SME2 currently uses his expertise and knowledge as a business offering to other companies in need of his services and sees himself as a consultant in the antenna sector.

Table 8: Collaboration between SMEs and other stakeholders.

Company code	Collaborations	Integrations	Possible outcomes
SME1	Other organizations	Shared resources, access to new network	New business opportunities. Cost sharing.
SME2	Other organizations	Contract services	Product development. Risk reduction.
SME3	Researchers	Co-development	Product development.
SME4	Municipalities, Researchers, Other organizations	Shared resources, Co-development, Contract services	New business opportunities. Cost sharing. Knowledge sharing. Service development.
SME5	Other organizations	Shared resources, Co-development	New business opportunities. Knowledge sharing.
SME6	Partners/ Customers	Co-development	Ensuring that products fit into core systems.

Table 8 highlights the different collaborations and future collaborations that were mentioned throughout the interviews. Since the companies have asked for anonymity the collaborations are mentioned in general terminology. As seen the interviewees had several different reasoning behind different types of collaboration, every SME in the upstream space industry and including one of the downstream entities was open to be a part of an external platform in the sense that they acknowledge their systems must fit into a core system, which aligns with the theories about complementary innovation and external platforms that were covered in the literature review.

4.4 Stakeholder involvement

All of the respondents from our interviewed SMEs have considered who their stakeholders are and in what way they affect their direct involvement and operations. Most of the interviewees identify similar sets of stakeholders for their organizations (see table 9 below). Customers seem in every case we studied to have the most power among the relative stakeholder relationships; in stakeholder literature it furthermore reflects on the power of so called definitive stakeholders.

The involvement from the stakeholders vary and as for their product the general direction points towards higher involvement as the delivered product or service needs to fit into a core system produced and controlled by the partner relative to the SMEs collaboration.

The interview from SME6 explains how the direction of their undertakings are affected:

"And Airbus, If you have the customer, they of course have a strict specification because the product needs to fit into their system [...] What exact performance is needed is to a large extent driven by the industrial customer.". (SME6)

While the likes of big organizations such as Airbus dictate the flow of business the vice president of SME6 adds further that it is a necessary way of conducting business.

"If we would have done it alone we would not have succeeded [...] you will not fit into their system if you do it on your own and they will not trust you." (SME6)

The respondent SME3 has an aligning view and states that their innovation process is directed by the stakeholders involvement:

"It does inhibit innovation, I would say. I am not objective in this, because I believe this is often how innovation is done". (SME3)

The respondent SME4 has also identified the necessity of collaborating with different stakeholders in order to be able to conduct business. They start their development process by asking researchers what needs and wants they have in order to tailor a solution that works for them rather than selling a generic product or service.

"We have a lot of contact with researchers to see if we can co-develop something together". (SME4)

The one company that has diverged somewhat from the generic business offering to the others is SME1 who is currently keeping the integrity of the business. The company still needs to adhere to what the market demands but is currently not influenced by stakeholders in their innovation process.

"We are bootstrapping [...] Once our first qualification is done and we have proven the interest in our business we need to re-evaluate if we want to continue bootstrapping or if we want to go out to the more commercial business". (SME1)

Meaning that the company may diverge from their current path of independence if the current business method does not prove to be economically viable, so while they want to be independent from outside influence it may not be a sustainable strategy. While they want to keep their integrity they also adhere to the market demands, they do however state that their offering is industry standard.

"We are working with industry standards, best practices and what we produce is a test report that our customers can read and be sure that we have adhered to best practices and standards". (SME1)

Table 9: Summary interviewee stakeholders.

Company	Stakeholders	Typology	Type of stakeholders	Comments
SME1	Potential employees, Consumers, Financial contributors, Universities	Discretionary stakeholder, Definitive stakeholder, Dominant stakeholder, Discretionary stakeholder,	Internal, External, External, External	Attracting the right potential employees are paramount. Customers currently have decision power. Contributors assert much influence while others are more passive. Contribution from universities is paramount.
SME2	Customers, Swedish space corporation, Universities	Definitive stakeholder, Dominant stakeholder, Dependent stakeholder	External, External, External	Customers are currently governmental, looking for other options. SSC is vital as they provide new business opportunities. Universities are highly important due to co-development strategies.
SME3	Big organisations, Investors, Employees, Incubator, Customers	Definitive stakeholder, Dormant stakeholder, Discretionary stakeholder, Non-stakeholder, Definitive stakeholder	External, External, Internal, External, External	Organisations dictate innovation direction and are potential future customers. Investors have no impact on the business other than wanting return on investment. Employees add legitimacy to the firm by innovating. Incubators moved from an active to passive position.

				Customers likewise dictate the innovation path for the company.
SME4	Researchers, Universities	Demanding stakeholder, Dependant stakeholder	External, External	The company wishes to enhance researchers to create a better society. Co-operates with universities to reach and influence researchers.
SME5	National customers, International customers, Community, Society	Definitive stakeholder, External stakeholder, Demanding stakeholder, Dependant stakeholder, Dangerous stakeholder	External, External, External, External	National customers act as teachers for SME5, showing them the way of their business. International customers show interest and could in the future be potential customers. SME5 wants to help regional communities. The trends in society dictates the direction for the company.
SME6	Owners, Customers & potential customers, Swedish space agency, European space agency	Dependent stakeholder, Definitive stakeholder, Discretionary stakeholder, Discretionary stakeholder	Internal, External, External, External	Owners have design decisions but are dependent on industrial partners. Customers decide how products should be designed to fit their core systems. SSA provides valuable industry information and gives innovation directions. ESA also proves valuable industry information and innovation direction.

The overall most common answer to who potentially holds the most power over firms, the general answer was customers, in particular when dealing with big industrial organizations. In one of the cases researchers were highlighted as an important part of their business due to the impact they have on the innovation process in the firm, saying that their development would not be where it was without constant lines of communication with the academic world.

4.5 Motivation

It is a difficult task to enter the space sector with a new idea, interviewees have established that it is a conservative market (SME6) where old relationships play a major role in establishing what organization should get opportunities to support the ecosystem (SME2). There is also a reluctance from private capital to be engaged due to the uncertainty surrounding the industry (SME6) whilst also public resources are limited and old legislation dictates the pace of investment (EXP3). But, still certain individuals and organizations have challenged all difficulties involved and created opportunities for themselves. There are several reasons as to why the companies interviewed have entered the space sector. The reasons range from essentially all motivational concepts, some of the interviewees have tendencies towards several while others are leaning towards one.

One main reason for several interviewees seems to be that it is a technologically interesting sector where companies can solve interesting problems, opposed to working for big corporations where individual freedom is far more restricted, while it does not establish why the space industry was chosen over another field it does explain a certain mindset that the individuals interviewed holds. SME1 frames it as follows:

"My motivation for starting SME1 was to step out of the established space industry and to be able to try my ideas without the extra hurdle of "it hasn't been done before". If you try to do something inside a company if you are not the owner or have critical decision power then you have to sell that idea in-house. And with space being a very conventional and conservative business that can be a real challenge.". (SME1)

That the space industry is conventional and difficult to get into is also confirmed by EXP2 who shares that sentiment and adds further by stating:

"How do you build something that flies in space?
Well, you need to have flown before. How do I get through the door the first time? Well, you do not, it is a cultural and procurement thing. That is not an obstacle for startups in other markets. Internet, mobile, tech for example. They do not exist there, they have gone away over the years and that is when we can see a rapid growth and it will happen in the space sector as well. But it's taken a little bit longer than other markets unfortunately."
(EXP2)

The statement suggests that the choice of industry is not rational in a business opportunity sense. It also adds the belief that the space sector has a bright future when and if the conservative nature of it is abolished and companies gain access to more opportunities than today. SME5 raises the point of space offering interesting problems to solve reflected further as:

"If you really boil it down, it is about solving interesting problems. I think that's actually what makes our clocks tick. It is much more about solving interesting problems and doing something that feels like a really good solution more than making a lot of money.". (SME5)

Which further raises the question about profitability within the sector. Only one interviewee states that working in the sector is significantly profitable for him and he is one of three that are involved in the upstream space sector suggesting that if one succeeds in making upstream applications for a major industrial organization then there is profit to be made. The other interviews revealed several different motivational factors that gravitized them into working in the space industry.

"If I wanted to make much money there are many other places I could go. Some coursemates went into finance and are working on creating complex mathematical models for the stock exchange and things like that and that would have been much more worth my while in terms of money but not as fun." (SME1)

"It was very exciting to work with communications and to work worldwide [...] it was quite profitable. That's why I started in this from the beginning. I like to build up things from the beginning and not just be a part. I like to develop." (SME2)

"I have a societal call because I believe that work that people don't have to do should not be done by humans. It should be automated, and humans can focus on the things we need to do. so that's one motivation. To automate society." (SME3)

"For me it was a long life decision actually, I always wanted to be my own boss [...] I worked within this industry for most of my life." (SME4)

The industry is no doubt interesting in a technical sense for the individuals interviewed and they tend to value problem solving, societal calls, the freedom of being the decision maker more than using their degrees in high profit industries or being employed by a large company with a high salary. EXP3 answers in line with what we found out from our SMEs interviewed regarding motivational factors to enter the space industry.

"Space is quite a hard sector to get into. Space has always interested a large part of the population and is perceived as a very high technology area. I think that if you have a good idea you would want to test it to the max and you choose space to do it. And if you can show that it works in space, then you can be quite sure that it will work in all other environments on earth. So I think that is one part with new space and with the app culture. It is also possible to use space data in ways that we never thought of before. And the time to market and the time to develop and get products accepted is much, much, much shorter,

which I think could be a driving factor. It's possible to make business in quite a short time."(EXP3)

It goes well in line with what EXP2 stated earlier:

"Many entrepreneurs, they really are into the technology and the space sector and solving some real business problems comes secondary to that. But that's usually the way it is with entrepreneurs." (EXP2)

EXP1 is in agreement and does not generally think people enter the market to make large profits:

"Maybe somewhere deep inside people entering the space industry want to become rich. But I think they are going to be disappointed. I mean just that space is not a moneymaker. They used to joke that the best way to make it a good space company is to make a good IT-company, become a millionaire and then start burning that money". (EXP1)

Table 10: Different motivational incentives for SMEs.

Company code	Motivation	Motivational concept	Comments
SME1	Individuality, knowing the industry	Risk-taking, Self-efficacy	Worked in the space sector for many years and decided he wanted to continue without the restraints of being in a big organization.
SME2	Knowing the industry, profit, excitement	Goal-setting	Worked in the industry for many years.
SME3	Societal call, business opportunity	Risk-taking; Need for achievement, Locus of control, Drive	Want to automate society. Saw an opportunity in the space sector with his innovation.
SME4	Individuality, Knowing the industry, societal call	Need for achievement, Locus of control	Worked in the industry since graduation but essentially wanted to be her own boss. Believes environmental causes are important.
SME5	Solving interesting problems	Egoistic passion	New to the space sector. Attended conferences out of interest for space and saw business opportunities to be made.
SME6	Profit, knowing the industry	Goal-setting, Independence	Been in the space industry since graduation. Saw a profitable business opportunity.

4.6 Upstream and Downstream space

We have established earlier that it is important to separate upstream and downstream processes but that they uniquely form according to the ecosystem in place. EXP2 explains that it has caused misunderstandings historically when not being clear of its distinction.

"We have learned that when you talk you need to separate upstream and downstream. When you talk space upstream then you can present rockets, space suits and requirements for lower earth orbit or geostationary. Then you can be in the space space framework on how you express things. But when you work on downstream applications, like talking about flying satellites or drones, then you need to be using normal tech startup marketing and wording. But with a little bit of the space marketing twist at the end. Do not combine the two because people are going to think space suits." (EXP2)

They work differently, have different actors and are concerned with different barriers for entry. The latter is far more reasonable for an SME to enter as the economical, trust, and reputation barriers are lower (EXP3). Most new organizations can afford to opt into for example google earth and create products and services with satellite data and get into the downstream ecosystem that way. The upstream space branches have significantly higher barriers related to economy, relationships, and trust which can be more difficult for an SME to successfully enter and succeed (EXP2, EXP3). The way to enter the upstream sector is by creating a product which can be tested and successfully applied in tough environments on earth and in space for many years to prove their longevity both as a product and as a company. No big corporations who are sending up rockets to space are willing to risk their operations with an unknown actor's product (EXP3).

The downstream branch has been historically overlooked by the government and the funding agencies who have overseen the space industry mainly focus on the upstream area due to a conservative view of the industry. Things are however changing, the expert interviewed within the funding agency stated that they in recent time have employed personnel whose sole purpose is to act within the downstream branch. The EXP3 states that it is paramount to make resources available to both streams of the industry as they coexist and one cannot exist without the other. The change has been acknowledged further by EXP2:

"'It needs to be reviewed (the space laws). Just one example is that the Swedish national space agency, up until two years back would only work with things going up. They did not want to bring down satellite data and start a company that uses images that are downstream. That is not us they said." (EXP2)

EXP1 adds to that further and states that he is seeing a change in the landscape:

"There is a change in everything related to space and these more agile new space solutions are going to come into even the big science missions." (EXP1)

All three experts are in agreement that the sector is changing or at the very least has to change, the new initiatives needs more autonomy than previously since it is changing at a always increasing pace and the old ways of doing things are too inefficient in today's landscape.

4.7 Outside-in, inside-out, and open-innovation

Most companies interviewed had in one way or another an outside-in approach to their innovation efforts. The inside-out sentiment was likewise a common way for the SMEs to make profit while spreading their knowledge to others who are in need. Open innovation was only identified in one of the cases but has had a significant upside to the business in question.

One interviewee who consistently worked in an open innovation setting is SME4. They have much contact with research in their business endeavors and have collaborated with both Stockholm university and Australian research organizations in order to co-develop services in line with an open innovation setting. They utilized researchers in order to find new markets, co-develop, while broadening their service offering to gain new knowledge. They state that even though they innovate through open innovation it mostly happens in-house. The second SME utilizing open innovation is SME5, who stated further that they have no problem to co-develop products with firms generally seen as competitors in order to satisfy a market need. SME5 further describes their innovation efforts as:

"We want to help our customers innovate. We think it as more of an evolutional way of innovating than a revolutional. We do not aim to be disruptive or think totally outside of the box. It's more like incremental improvements and always working to make it a little bit better every time" [...] Everybody stands to gain from that, both us and the competitors and most of all the customer. If they can have a combined competence to deliver the best project, then that is a win, win, I think". (SME5)

More specifically our interviewee suggested that there are potential benefits to be achieved when collaborating in an open innovation setting which enhances competence, skills and knowledge for all participants that were previously non-existent.

SME3 started his company based on a master book he wrote in university:

"It all began when I came up with an idea about how to create a very lightweight but precise robot arm. I invented the technology using photonics and integrated a sensor system in the arm and started to write a description of the system [...] I am a Master of engineering physics so I in my studies could work at my own company with my ideas and still get credits. I received a patent and it began from there, through many iterations I verified the technology and developed the core systems and built up a team around myself, at most we were 25 people working on the project." (SME3)

Today he also applies an outside-in innovation sentiment by hiring consultants to design specific parts of his innovation that the firm is currently lacking knowledge to do themselves. They furthermore offer their expertise to other firms on the market and hence adopt both an inside-out and outside-in innovation setting for the development of both their own products but also in coordination with the whole ecosystem. SME1 uses an outside-in approach for development but not regarding the services that he offers but rather on a higher business level.

"Right now we take in consultants to develop our processes, plans and infrastructure" *(SME1).*

EXP1 explains that he has seen several spin-offs from universities which have become successful companies. One of the successes being a world leader in synthetic aperture radar. Another example is a software company specializing in creating software for cube satellites. He believes this is an effect of individuals not enjoying the slow pace of universities and the limits within and are really ambitious to "fly high".

The origin of the company somewhat determines whether or not a company has an inside-out, outside-in type of an open innovation approach to their development. Most companies we interviewed have worked within the industry for several years and created their companies out of a desire to be their own boss and work in an industry they acknowledge. They tend to value the knowledge they can gain from outside sources as a fundament for their overall strategic business decisions.

4.8 Opportunity identification

EXP1 states that academia is crucial for presenting opportunities for SMEs in the space industry. If there were no research institutes the downstream companies would simply not exist. An example of areas in need of innovation, which can be done by smaller companies are accordingly stated as:

"We realized there was a problem with asteroid science, let us say that every mission cost 100,000,000 and one mission can only visit one site [...] but in order to really understand asteroids we need tens and hundreds of data points and we cannot do that with the current method, so in order to increase our knowledge about them we have to innovate". (EXP1)

He furthermore states:

"I think earth observation companies can be big. They can become really big profitable companies". (EXP1)

SME4 highlighted the increased availability of free data that can be used to make services has sparked an interest in both supply and demand in earth observation products compared to a few years ago. SME4 expressed the following:

"So I think the free data and in combination with these platforms, they are the reason why it eventually has started to be a demand for satellite remote sensing products. Not only for the largest, the governmental organization, but also for smaller". (SME4)

There does not seem like there is a general lack of opportunities to pursue in the sector but instead EXP2 further explains that there are many opportunities for smaller companies in Sweden to enter the space industry, but the problem is the lack of support, resources and capital to let them successfully establish themselves.

SME6 has an aligning view on the problem but is specific to the public budget and says:

"I would say you cannot pull in more people with the current budget, but of course you can make changes. So yes, there should be room for new ones but then of course you need to decrease investment towards today's players that are maybe not having success or even having a possibility for success". (SME6)

He sees particular opportunities in the northern region of Sweden due to Esrange space center which can grant global interest in the region if the offer is packaged correctly and hence business possibilities for many local entrepreneurs will develop and emerge.

SME4 states that Sweden has the potential to create a wide variety of services and technological solutions that can be capitalized to further improve the current ecosystem to push Sweden to new space initiatives that will benefit society, industry, and small businesses. Through further analysis we can identify that new opportunities are and will be created with technological development, and the more these tools and technologies are utilized by the public sector it creates further influence and opportunities for newcomers in new ways of accomplishing goals, and objectives in the space industry. Such developments and objectives to tackle and to ensure such a transformation are within reach of Sweden, and consequently will fundamentally grow the space sector. SME4 believes there is room for much needed growth in the space sector in Sweden, and that it has been relatively slow to develop a new frame of doing things in the space sector due to the attachment of old solutions.

SME4 specifies further:

"It's not like the industry will just explode like this because people, they sit with their old solutions and the old way of doing things and it takes quite a time and to get accustomed to thinking in new." (SME4)

SME5 identifies that there is substantial growth occurring in the space sector, and that there is an ever-increasing demand from the Swedish government to ensure productivity and growth in the space sector, but that there remains a lack of awareness which impacts the amount of opportunities that reaches potential adopters.

SME5 further states:

"We actually did not know how relevant Esrange was. I knew that there was a site called Esrange and that they were launching balloons and small rockets. I was not ever aware that it was actually a very attractive test site for the European or maybe global market. I had no idea that it was growing like that, so it wasn't on the map at all." (SME5)

Sweden has a great opportunity to be a leader in the region for the space industry with launch sites that are not available anywhere else in the region.

SME6 describes that downstream is driven by technology developments that were unattainable in the past, and instead of governments investing into upstream which is highly relative to old space mentality they are actively looking to purchase services even if some of these services were already existing thirty years ago.

SME6 distinguishes further:

"New space is of course technology driven. That technology has evolved, and now certain solutions are possible by using for instance, much smaller satellites to provide a good service that was not possible before." (SME6)

Being in a quick evolving industry that is in many ways controlled by industrial giants and governmental decisions proves to impact the opportunity creation negatively, the technology has come a long way while rules and legislations are lagging behind.

EXP2 indicates there is room for improvement for Sweden to start utilizing space data from satellites to market and exploit downstream business opportunities, whereby he feels is lacking considerably due to a loss of interest in comparison to their existing objectives for research intensive programs and projects. While also the current legislation model and directives need to be improved to put into consideration a set of clear responsibilities to be developed and administered by certain agencies to supplement growth in the new space sector. Furthermore, EXP2 believes that the US model that is currently succeeding in the global market should be adopted and considered, but whether it will be considered or not will depend solely on the government's intentions and initiatives.

EXP2 clarifies further:

"Should it be adopted? Yes. Will it work? Yeah, of course. Can it ever be adopted? I do not know. Probably too many bureaucrats worrying about the EU procurement legal stuff and not just testing it unfortunately." (EXP2)

Table 11: Summary of empirical findings.

Theme	Takeaway	Comments
New Space in Sweden	Sweden is lagging compared to more efficient nations.	The entirety of the respondents agrees that efforts in the national/regional market are inefficient compared to other nations. But there are things happening, budget allocation has shifted, budget size has increased; the current space legislation is being reviewed.
Collaboration in New Space	Establish collaboration	The space industry is highly technical, and the majority of our interviewees say that different collaborations have led to an increase in turnover and innovation in their respective organization (and sector as a whole).
Stakeholder impact	Inhibits innovation but is often necessary	Stakeholders are paramount in terms of general knowledge for our interviewees. Stakeholder involvement from customers who have the biggest involvement is different depending on if the business offering is tied to upstream or downstream. In upstream it is a few big actors who have a big impact on the innovation process in an SME.
Motivation to enter	Interesting industry, problem solving, societal call	The motivational reasoning behind entering the sector is both according to the interviewees and experts related to soft factors such as wanting to make society better or that it is an interesting environment.
Upstream and downstream	Governmental oversight of downstream space	The more traditional, conservative upstream have been the focus of governmental efforts, which according to all three experts must change for efficiency reasons.
Creating an opportunity	Importance of research institutes. Increase of available data. The rocket range Esrange.	Over time the amount of free, available downstream data has increased and that can be used to create business offerings. Sweden has a unique opportunity due to having the only rocket range in Europe.

5- Analysis & Discussion

In the following chapter we discuss our findings with foundation in our empirical findings and analysis, furthermore present our inspired model and conclusive remarks. There is an existential debate on whether old space or new space has what it takes to exponentially shift the current paradigm in place.

Whether the traditional ways of incorporating and executing funding, or simply the access to resources for new entrants in a capital-intensive, relationship and network progressive industry. There lies a crucial aspect of forward-thinking, and excessive testing of materials, products, and systems which lay forth and encompass innovation on both sides of the paradigm. While regulations, and traditional systems prided by old space limit accessibility that therefore reduce new entrants to exercise their innovative abilities. New space brings in turn new possibilities that were not simply imagined a decade ago, and these possibilities have intertwined with old space and the traditional way of doing things, furthermore evolving.

Our research shows that there is a reliant nature between both new space and old space, but that this shift is slower in some regions than others. Furthermore, the capital requirements to join such an industry are further becoming increasingly prevalent due to shifts in laws, rules, and regulations to embrace small and medium-sized businesses and new entrants in Sweden. Such a system can only be viable if the infrastructure is in place, thus old space is highly vital for the assimilation of resources and platform synergetic effects that embrace new entrants. Though such relationships are critical to the establishment of cooperative endeavours between large traditional players in the space industry, and new small and medium sized entrants, there lies the importance of governmental associations and collaborations to ensure the overall success of the space industry.

The Swedish government according to our research has progressively and adaptively shifted to embrace new space initiatives and the correct system is in place. Though our participants in our research visualize better circumstances and situational factors that increase their opportunistic goals, there has been a significant shift to further embracing and ensuring funding is in place to further balance the scale. There are higher investments currently in place that further enhance old space players and large organizations, but according to our research this is due to long lead times, and infrastructure requirements for a space industry to be successful. As without the required infrastructure, simply correlated, the space industry will lack in progress, therefore, old players and new players will simply decrease in numbers.

The importance of shifting from old space and new space to be co-existent heavily relies on the system in place, and whether old space players and large organizations will cooperate with new entrants within the ecosystem. According to our research it has shown this can be possible, but it takes time and a great innovative feat that further enhances the small and medium sized businesses opportunity to work with the larger organizations. Furthermore, the exchange of knowledge and the degree of openness of organizations plays a key factor in innovation. While the establishments in place that promote knowledge and research prospects are essential to the innovation source, network, and destination. Universities play a key factor in an open network research environment that prospers innovation and are relatively fluid within several platforms whereby financial concerns and gains are not critical for coordinative knowledge sharing and expansion of innovation. Furthermore, funding is highly critical when it comes to open research environments especially in the space industry for small and medium sized businesses, universities, and let alone for countries to collaborate and pursue space initiatives and endeavours. Thus it is highly important for a country pursuing or strengthening their relative position in the space industry to consider a thorough evaluation of accessible funding due to the relationship funding has with the nature of performance.

5.1 Opportunity forming

There seems to be a few factors present at large which contributes to certain companies wanting to enter the space industry. Our findings suggest that there are few possibilities to make significant profits within the sectors which would be the most obvious reason when choosing what opportunities to pursue. Most individuals pursuing a career in space related undertakings rather seem to be a dreamer, not focusing on high profitability. They have a general interest in technology, wanting to solve problems that have not yet been tackled by our societies and have an urge to test their products in the harshest environment currently accessible. They see an opportunity to advance society in different ways, by doing what no other person or company has done before them. They are individuals who see opportunities which essentially are non-financial beyond making enough profit to make a living and advance their undertakings and rather have a quench for wisdom to dampen and a will to do, to them, something that matters depending on their respective environments (Meyer & Rowan, 1991, p. 340).

There are as mentioned several times a general lack of interest in investing in the sector for SMEs, both in the public and private sector, which we believe is a contributing factor as to why the swedish space initiatives are lagging behind other geographical areas in the world. This is identified by several experts and companies alike to be an area which is improving in present time, if it is enough to pull more actors into the industry is unclear, but our findings would suggest it would. Being constrained by resources surely inhibs innovative undertakings within the space sector, organizations and entrepreneurs would be far more likely to put their efforts in the industry if they knew there would be more financial backing to claim (Bruton et al., 2010, p. 422) . There are also indicators that suggest a more general reasoning behind pursuing space related opportunities, a more rational take where the individual have built up knowledge and networks related to the industry and want to pursue them without the restraints of being bound by academia or large companies which could dictate both the pace and direction that their ideas should form (DiMaggio & Powell, 1991, p. 130) . A reason we believe to be common regardless of the industry an individual chooses to pursue opportunities.

In both scenarios the individuals finding opportunities related to space are in many ways abstract thinkers who want to pursue what they feel is important and want to do it the way that they perceive is the way to do it. They want as few boundaries as possible to freely test their ideas and products. For some space is an active choice due to it being the toughest challenge humanity is currently facing and for others opportunities have formed over time and ends up being the only natural place where they could have ended up. There are no doubt many talented people with the necessary skills and knowledge to seek solutions in the space industry that are currently not within the ecosystem.

The marketing efforts to pull actors into the industry is slim, there are governmental organisations who aim to inform companies with the right competences into the industry, one of them was part of this book, but the majority of the interviewees found their way into the space industry by working on big corporations and eventually leaving to create something of their own. This book is not in marketing, but one example of approach could be the narrative of "the most difficult environment possible" used by funding organisations and other governmental organisations to attract people who seek those types of challenges and hence increase the absolute number of actors, as our findings suggest it is an attractive approach to the problems seeking solutions in the industry.

5.2 Stakeholder theory

Our book suggests that many SMEs who are pursuing opportunities within the space industry are in many ways forced to take a certain path, creating a paradox in relation to how the respondents answer questions related to opportunity forming, where they often choose a path away from big corporations to be able to be more free. More often than not the companies interviewed are delivering parts for a system or a sub-system and they hence need to follow strict manufacturing procedures to stay relevant in the ecosystem. Such subsystems and interfaces are a vital form and common structure from which a company operating as a platform leader can efficiently develop and produce a family of products (Gawer and Cusumano, 2014, p. 419). Thus further validating the heavy reliance of small and medium sized businesses on large organizations that are industry leaders for accessibility and entrance to operate in the Swedish space industry.

Our book suggests the rigid structure where only a few stakeholders truly decides the direction of the industry greatly impacts the innovativeness for the ecosystem as a whole. Companies that otherwise would let their creativity run free in order to break status quo are rather forced to be a small cog in a large machinery, financed by organizations who seem not interested in what the small and medium organizations could contribute if they were supported in a better way by their government stakeholders. This is heavily related to the public funding which today is focusing on the big traditional enterprises, making it difficult for smaller entrants to develop and innovate. Regarding the private funding it is dependent on the perception of the space industry being filled with uncertainty and risk. This is historically true, but today there are massive opportunities waiting to be tapped, especially regarding the earth observation bransch of the space industry. Companies can with little investment, thanks to new platforms such as google earth, reach a profitable market offering. It is according to us up to governmental agencies, such as incubators to convince the private capital that there is a growth market that they have overlooked in terms of investing. The funding is in both regards shifting towards a more SME friendly environment and we expect this will have a positive outlook on the amount of entrants in the near future. The funding organ has simply reached the same conclusion as us. Entrepreneurs do not want to be tied to a big organisation to supply them with subsystems but want to innovate more freely. Further increasing the funding towards smaller companies should attract more entrants, meaning that Sweden is on the right track in that sense.

In order to foster a true innovative nature in the Swedish space industry there needs to be a shift of balance of power in the stakeholder relationships currently present. The space industry is innovative in its very nature, but it is simultaneously part of a system where entities seek the highest possible profits in the shortest possible time. As long as those entities are by most actors considered definitive stakeholders there will not be a change in how the ecosystem functions, as they are driven by profit and calculated risks, which again, inhibits innovation which many of the small and medium organization truly seeks but within the current stakeholder structure cannot achieve. Managers are essential in the development of relationships, inspiring their stakeholders, and in the creation of communities whereby everyone's involvement strives to give their best in delivering the most potential value the firm promises (Freeman et al., 2004, p. 364). Definitive stakeholder's hold all three attributes in the stakeholder identification process which consist of power, legitimacy, and urgency thus imploring dominant control among their stakeholder relationships (Mitchell et al., 1997, p. 878).

Furthermore, if the power transitions away from the big organizations the universities and academia will play a bigger role in the stakeholder relationships with the small and medium organizations. Therefore, a party identified in terms of a relationship has power, to the extent it has or can gain access to coercive, utilitarian, or normative means, to impose its will in the relationship (Mitchell et al., 1997, p. 865). If the smaller and medium sized organization are more free to test their ideas without being tied to a big organization the ideas generated in academia will be more accessible and relevant, which in turn would push the Swedish industry to be more relevant in the global scene, creating more organizations which are in the forefront of innovation. Hence making other stakeholders outside of Sweden who currently have little interest in the Swedish small and medium sized companies to rely on them for creating the technology that they would need in their business ventures, creating a spiral of positive effects for our national industry. Essentially, the current stakeholder system works if the goal is to survive and create some return, but it will not increase the global relevance when there are more agile ecosystems in other countries.

5.3 Platform theory

Platforms in the Swedish space industry do according to our book play a big role. The choice of business direction does dictate in what way, it ranges from simple to absolutely paramount to business success. An example of where a company is very reliant on a platform leader and the access to their technology is an organization utilizing google earth, not being able to access that platform would render the business idea impossible to execute. Other examples found in the book is the reliance on platform leaders to review and test products before companies are able to get them to market. If the relationship with such platform leaders are in bad shape it could potentially deny the company to do business, further limiting opportunities for small and medium sized businesses. Stakeholder relationships and identification measures assist managers to map the legitimacy of stakeholders and whether to be salient and sensitized to the moral implications of their actions with respect to each stakeholder (Mitchell et al., 1997, p. 880). However, there are also more positive examples within platform theory, some businesses have found synergies collaborating with others to strengthen their position on the market. These companies are integrated with each other and are relying on each other for business success in a harmonic structure. But the other side of the spectrum is ever present for some of our interviewees, being integrated with platform leaders who are solely interested in monetary return or are there to regulate and review products and services can prove to be a ponderous ordeal. Platform owners and leaders are further described as a firm that owns a core element of the technological system and whereby it defines its forward evolution (Gawer and Henderson, 2007, p. 4).

Entering the space industry is considered difficult for a variety of reasons, which is also true when considering platform theory. There are indicators in our book which suggest that it can be smoothed out by finding the right partners within the platform where the idea is relevant. They can provide knowledge, support, co-creation, and testing of the products and services which can greatly improve the relevance of the organization and reduce the time to market, which is suggested by the book to be important as one of the main problems to go into the industry is the time to find returns; smaller and medium sized companies simply cannot survive several years without seeing some return on their ideas. Within the book the companies who have functioning platform integration have faster time to market, more direction and an easier time to translate their ideas into revenue, which is preponderant for long time success. The trajectory is simply better for companies who have integrated themselves into a relevant platform. Major beneficial implications for a firm with access to a platform network and supply chain is the enhancement of their external capabilities in seeking more innovative or less expensive components and technologies (Gawer and Cusamano, 2014, p. 419). Companies who have not found that integration seem to run into far more problems than those who have integrated.

Our book suggests that finding the right platform partner is one factor which impacts the likelihood of success the most in the Swedish space industry. Business is the mutual benefits a deal can attain so that suppliers, customers, employees, communities, managers, and shareholders all win continuously and predominantly over time (Freeman et al., 2004, p. 365). Internal platform capabilities can also be optimized by large organizations and platform leaders to further enhance several potential benefits that include savings in fixed costs; efficiency gains in product development in the ability to produce and extend their product lines with limited resources; while furthering flexibility in product features and designs serving and attracting several customer specifications (Gawer and Cusamano, 2014, p. 419). Thus further describing the relevant direction and strategy by large organizations in the space industry, and their selective composition in choice of partnerships, and business endeavors.

5.4 Complementary theory

A noticeable trend among interviewees was that they all had a positive view on different collaborations. Two of our interviewees are completely dependent on collaboration, another two interviewees are collaborating but are not as dependent but still choose to collaborate, one interviewee is planning to collaborate and one interviewee has a pragmatic view on partnership. Our book hence suggests that it is difficult to be in a "know it all" position in the space industry where conducting business in a vacuum independent of other companies seems close to impossible. Companies tend to be dependent on new information reaching their organizations from academia to be able to advance their own offerings on the market and seldom try to do the research themselves. Conferences tend to be another forum to find the current trend and to get a feel for the future market need for several companies. The companies interviewed have a calculated approach, rather than guessing what customers will be needing from them.

The companies are often suppliers for bigger organizations and are delivering sub-systems or specific products for those and having a dialog with them hence becomes paramount in order to be relevant. Demand for the platform is derived from the demand for the overall system (Gawer and Henderson, 2007, p. 4). Adding on, the book suggests that there are often requirements for the products and services that fall far from the company's core competencies, further building the need to have collaborations in terms of testing and consulting about quality requirements. The trend for collaborating creates an open setting in the space industry, at least in terms of smaller and medium sized organizations. Knowledge flows both ways, enabling these companies to be in the industry at all, without the perceived or actual openness they experience it is unlikely they would be able to be a part of the ecosystem. This is a tendency that might shift as the Swedish space industry is influenced by other regions or evolves. But as it stands, the big corporations are dependent on smaller companies for sub-systems so they need to be open about their designs and functions and smaller companies are dependent on them and on academia for knowledge and know-how information. Most interviewees base their undertakings on research coming from academia, making academia a paramount cog in the ecosystem.

In moments whereby a monopolist cannot duplicate a third-party complementors' innovation at a reasonable cost, the large organization will have strong incentives for commitment (Gawer and Henderson, 2007, p. 5). But this predicament can also be connected to the relatively low funding which is reaching the smaller firms, if that increases they may be able to produce more complete systems and conduct testing themselves, and over time have the capabilities for economies of scale moving companies further from each other entailing price sensitive competition. If a monopolist's or larger organization's incentive in engagement post price battles is sufficiently strong enough, complementors may have no incentive in engagement for innovation at all (Gawer and Henderson, 2007, p. 5). Thus such predicaments in complementary innovative practices highly rely on cross functional and beneficial outcomes for all actors within the platform and ecosystem to ensure mutual success and growth.

5.5 Proposed framework

In this final chapter of discussion/conclusions we present our proposed framework on how we believe the space ecosystem could, with efforts, transform into an environment that is more attractive for SMEs to enter. The framework is divided into three essential parts of the process of innovation, pre-innovation (planning and information gathering), innovation (product and service development) and post-innovation (adaptation, future collaborations and opportunity creation). The creation of these themes are done to give us a more clear structure of the process of innovation regarding the space industry as well as to give the reader a better understanding of the different mechanisms at work and pinpoint exactly which ones we propose need to change. The framework is loosely based on (Granstrand & Holgersson, 2020, p. 7) *Illustration of the innovation ecosystem* due to it establishing what is part of an innovation ecosystem and (Peeters, 2018, p. 188) *Transfer mechanism from public to private space activities* since it is more specific to our book than what Granstrand & Holgersson, 2020 provide for us and goes into how, in general, the industry has gone from old space to new space. Certain inspiration also comes from (Valkokari, 2015, p. 21) *Business, Innovation, and Knowledge Ecosystems: How they differ and how to survive and thrive within them* as it provides us with deeper understanding on how to be successful in different ecosystems. Together with our empirical findings the three models support how and why certain ecosystems thrive and let us suggest actions to improve the current system.

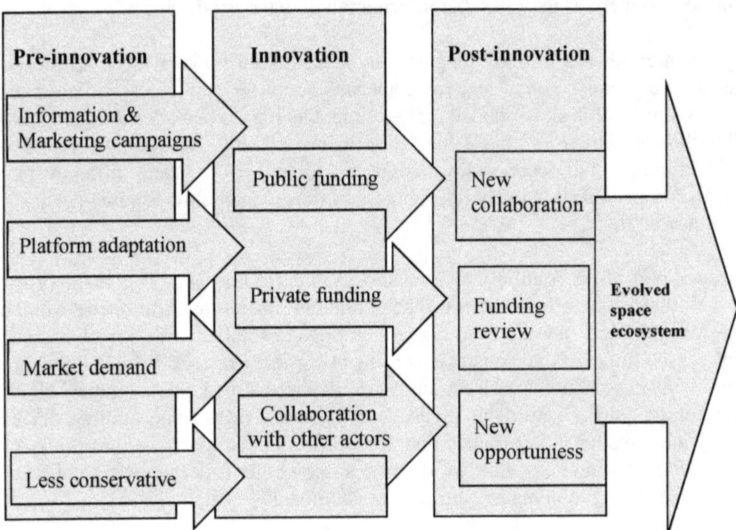

Figure 9: Innovation framework for evolved space ecosystems

The first part of the framework, pre-innovation relates to how our interviewees perceive how opportunities are seized, as information on what they wish could be better than the current ways and our own interpretations. Regarding information & marketing campaigns there seems to be a general misconception in the industry, the old space mentality is still strongly clinging on to the market, creating the view that it is not a viable industry to enter for most. The campaigns should be directed towards individuals and organisations currently not in the ecosystem but who have the ability to strengthen the ecosystem, they need the correct and updated view of what is happening and what will happen in the future so they can enter and contribute to themselves and the current actors.

The information and marketing efforts should originate from governmental institutions, for example incubators who have close relations with private investors and the relevant companies, matching those together. Next is platform adaptation, a fairly new and evolving way to conduct business in the space industry is through different platforms such as google earth and amazon who provide services that can cheaply and easily be adopted for certain companies. Not only do they create whole new business opportunities but they significantly decrease the required level of investment.

These platforms can serve as inspiration for both current actors as well as future ones and should be incorporated into governmental efforts to inform relevant parties on how they can be used and for what. Furthermore it is up to the platform providers to keep developing them to ensure a steady stream of new entrants. Market demand is heavily related to what was said regarding platforms, with information about the possibilities that the platforms can offer they will have a more clear idea on what they can request from the ecosystem. As of now, it is up to the actors of the ecosystem to offer their products and services to them as they lack the information on what is possible.

Lastly our interviewees strongly suggest that the big organisations which take most of the budget are very conservative and reluctant and rarely let new actors collaborate with them. It is a difficult task to change that sentiment as it goes back decades and it is quite possible the one dimension that is hard to influence directly, we do however believe as the shift in budget will decrease for them it will force them to take a different stance to be able to compete, so it is likely just a matter of time if the public funding agency keeps on their current track.

The second part of the framework, innovation, relates to how the firms make or want to make their innovation efforts happen. The public funding has as mentioned several times seen a shift in budget moving from big corporations to smaller firms, which is a positive outlook for the future of the industry according to our interviewees. While public funding, or funding in general comes with certain restraints such as spending tax money efficiently or investors requiring return on investment, it also creates opportunities. SMEs are generally more resource constrained than their bigger counterparts and have difficulty to attend relevant conferences, survive, develop among other aspects, more public funding will decrease those problems and attract more companies to the industry.

The private funding, in addition to the space industry is labeled as a high growth sector by governmental organisations such as Vinnova, as with AI and med tech, it will act as an eye opener for private investors and attract more resources to the ecosystem that way. Lastly collaborations among different private organisations are important, as it is difficult to be in a "know it all"-position regarding space undertakings and the fields within can be narrow. Our interviewees have identified this fact and embrace it to the best of their abilities and states that much of their innovation efforts happen that way. The space conferences in Sweden are small and underdeveloped when compared to other countries, which is an area of improvement that governmental organisations need to improve to attract more actors. Furthermore, academia plays an important role and can act as a catalyst for SMEs, not only can they collaborate together in a more elaborate manner but academia can also express what they will need in the future, sparking new opportunities. It is again up to public organisations to create an atmosphere where these two entities can meet and exchange ideas which could create new.

In the third part of the framework, post-innovation, is the outcome of the innovation process. The new collaborations aspect regards the desired outcome of innovations by SMEs, be it collaborations with SMEs, large corporations, academia, governmental organisations with different levels of openness and geographical boundaries it should always be the desired outcome after an innovation iteration to seek collaborations, it increases the absolute number of actors in the ecosystem. Making it bigger and hence creating more opportunities, which is the effect we seek. The interviewees tend to think along these lines, which is a good start but the rest of the ecosystem needs to follow, special effort should, as touched on in section one of the framework, be with platform leaders who in many ways drive the innovation today and it hence needs to start there and radiate elsewhere.

Next aspect to take into consideration is funding review. As of now almost all capital in the industry comes from the public sector, meaning it is tax money. It is important that there is a good review system in place when investing in new space technology, as good public image and perception is a must. Aligned public images and perceptions means that more resources can be invested, making it easier for new entrants to be adopted by the ecosystem. Furthermore it ensures that the investments are efficient and can be directed to companies showing promise. It is a positive spiral, important to the evolution of the ecosystem. Leading us to the last part which is new opportunities. When an innovation has reached the market it is important it is sufficiently adopted by the appropriate actors. It can be done through 'matchmaking' services provided by for example Ignite. It mostly relates to platform leaders who currently are reluctant to adopt most new innovations, but also the market as a whole.

If and when it is achieved sufficiently it can spark new collaborations, create more market demand and ensure the survivability of the innovation. Only through understanding that the provider can move back to the pre-innovation stage and identify other market demands which can be sought to further increase the size of the sector and in turn attract new companies by showing profitability they achieved.

6 Conclusions and Implications

In the following chapter, we start by sharing our conclusive insights with the presentation of our concluding remarks. In our second subchapter we represent our general contribution while fully emphasizing our theoretical, managerial, and societal implications. Lastly, we shed light and describe the limitations on our research while providing direction for possible areas for extended future research.

6.1 Answer to Research Question:

Within this subchapter, we present our conclusive insights and provide a direct answer to our research question with the presentation of our concluding remarks.

The main purpose of the book has been to answer the question:

"How do new entrants in the high technology space industry navigate the transition from old space to new space?"

By using our theoretical framework combined with the insights from our empirical findings we have critically analyzed, and provided a clear framework on how new entrants have adapted to the landscape within the space industry. Our findings show that in order to navigate within the space industry, firms currently have been directed by larger strategic and monetary goals set by large platform leaders. These organizations dictate the flow of innovation and overall design choices in the industry as a direct result, leading to newer smaller entrants becoming a subsystem of a larger system. The situation has in part been created by the government's resource allocation model, where the platform leaders are the only ones contending for public funding and partly by being in a dominant position since the earliest days of space exploration and development whereby they have stayed in control over the ecosystem until present time.

Moreover, our book shows that small to medium sized enterprises do not pursue space activities for large profits but rather to advance society and to challenge themselves in cutting edge technological challenges, meaning there is a dissonance between their identity and what the sector can offer. As identified in our research, a company's core reason for existing is to turn a profit for investors, smaller firms have had limited choices in terms of innovative, strategic, and creative freedom. Furthermore, they have had to disregard parts of their desired identity in order to satisfy the rigid nature of the sector. In order to sustain and remain, firms have had to follow suit and not be too disruptive to the industry leaders unless they are innovating new technologies that would enable the platform leaders. Our book identified that the ongoing shift in budget allocations from public funding will help change the dynamics and trends of the industry and render it more open and dynamic in nature. With more of the public budget being allocated to smaller firms they in turn have less constraints, and have the required funding to be associated in a capital intensive industry, which is vital for their inclusion to the space sector.

Moreover, old structures are being replaced by new, allowing entrants to more freely follow their innovative nature rather than being a sub-system of a larger system.

The process to enter and act within the industry has historically heavily relied on past relationships and knowledge. Without those factors, firms are unlikely to succeed due to the lack of power and infrastructure capabilities they require to be successful in the space sector. Relationships in the sector have been formed between platform leaders and those who they deem considered trustworthy and reputable. In order to enter the ecosystem, firms need to have the desired reputation and trust considered appropriate by platform leaders as they have complete control and dominance in the ecosystem.

Firms can to a certain extent disregard the trust of the platform leaders in favor of more dynamic relationships due to new services and products being more freely available.

Without the knowledge and the required infrastructure needed to succeed in the space sector it has proven difficult as it is closely guarded by institutions and organizations with close tied networks, making it difficult for new actors to enter.

As the industry and the whole world is shifting to a more open, fluid nature, the trust and collaboration between smaller actors becomes more relevant. Relationships, networks, and collaborative efforts can be created between them without interference from the larger more dominant actors. In turn that mindset increases the absolute number of potential partners in the ecosystem making it easier for entrants to pursue their core capabilities and innovations.

Predominately, knowledge is increasingly becoming more freely available, which forms new opportunities for new entrants; an aspect highlighted in our empirical findings whereby academia is suggested to create both knowledge and partnerships.

The availability of tools and readily available resources that have evolved and became the norm in society can be utilized by possessing knowledge. Fundamentally, shifting the power dynamics away from platform leaders. In turn creating a sector where small and medium sized enterprises can more freely create their desired path. As this shift further transcends from old space to new space mentalities, further devolpments will be based on coordinated efforts, interconnected network relationships, and cross collaborative integration without prejudice.

Therefore, ultimately leading to a new phase of evolution with a core objective towards a more unified space industry based on furthering innovation without discrimination.

6.2 Contributions and Implications

In this following section we state and discuss our research's general contribution, and its overall benefits in the implementation and advancement of innovative ecosystems in the high technology space industry.

General Contributions

First, by conducting our exploratory book, we further analyzed Sweden's current situation by gathering information and data from the ecosystem's participants, and the data collected contributes to improving Sweden's space industry. Second, the space industry is always transforming and evolving, and we show that our book contributes to the significance of the space industry and assists in the identification of opportunities ready to be exploited by the Swedish space industry, furthering Sweden as a potential future leader in the global space race. Third, a distinct overview and analysis on the difficulties faced by new entrants mainly small and medium sized businesses in the high technology space industry. Lastly, our research contributes further on the analysis, identification, and explanation of transitory factors and changes that have existed and evolved through a thorough retrospective analysis between old and new space.

Moreover these differences between the two sides of the spectrum are expressed and characterized to contribute foundational support for new entrants navigating such a transition in the high technology space industry.

Theoretical Implications

This next section states and describes our theoretical implications while highlighting our evaluative process.

First, by the utilization of our selected theories we aimed to further understand how relevant each of the theories are to any innovative ecosystem (Boote & Beile, 2005, p. 3). Second, by implementing a stakeholder's view on ecosystems and industry platforms we further understand motivational and influencing factors that identify the relationships for co-creation, co-development, and the formation of opportunities (Minoja, 2012, p. 71)(Mitchell et al., 1997, p. 869). Third, our research adds value to previous academic literature, by analyzing ecosystems from a new and fresh perspective through the combination of our selected theoretical framework and evaluations. Furthermore, in addition our research further contributes a unique perspective on the creation, implementation, and maintenance of networks, product offerings, collaborations, and strategic partnerships in a highly competitive and resource-dependent industry (Gawer and Cusumano, 2014, p. 1)(Valkokari, 2015, p. 17).

Fourth, the theories we used further exemplified such relationships and that ecosystems follow such a pattern in attaining network synergy, and cooperative models that are based on mutual goals and objectives (Christensen and Overdorf, 2000). Every entity in the ecosystem is an actor and each actor has their motivational and influencing indicators based on their relative stance and perspective on their stakeholder relationships (Mitchell et al., 1997, p.863).

Fifth, our research contributes further by highlighting and emphasizing the barriers and lack of power small and medium sized businesses have in comparison to the big players in the industry due to resource constraints, a lack of infrastructure capacity, and selective budget allocations in a capital intensive industry (Mitchell et al., 1997, p. 865) .

Lastly, our book sheds light on the current ecosystem that exists for the Swedish space industry and the data we gathered and analyzed shows opportunities to further strengthen the current ecosystem (Boote & Beile, 2005, p. 4).

Managerial Implications

Following, our corresponding value of our managerial implications, and by shedding light on the current ecosystem and the perspectives gathered from small to medium sized organizations and experts within the space industry.

Managers and new entrants will be able to overcome barriers that are identified in the book. The findings from our research dictate and exemplify what small to medium sized companies as well as experts have highlighted are the factors and expectations industry leaders require in the present time to participate in the ecosystem. Interviewees emphasize the importance of becoming a subsystem to a larger system within the industry, which on one hand creates opportunities to survive as a company financially.

But on the other impares the innovative and curious nature to which these companies and managers consider to be nested in their personal and organizational identities. New entrants and managers can take these built in industry expectations into consideration when forming their offering to the sector and hence build a more complete ground for decision making, and formation of opportunities. Furthermore, the interviewees have highlighted how the evolution of the industry will likely transform, therefore helping managers build more long term business strategies, and networking possibilities. They can in short with our findings consider their capabilities both according to the present situation as a necessity of being a sub system of a larger system controlled by industry leaders while preparing for a more open future.

Our interviewees underline their dreams and aspirations to become pathfinders of improving current technologies and finding new solutions to old problems.

As this book addressed; the current state of the industry is still relatively dominated by big industry players, it is clear that the future will and is becoming fundamentally different.

Changes to the industry will open doors both for current and future managers to be more in line with their perceived identities which will attract a new ideological shift in perspectives among managers. Both current and future companies will be more financially free due to the market being seen as a growth industry while also an increase in share of the allocated public budget spending to a larger extent will allow managers to decide whether to conform to the system or to create products and services more independently.

In conclusion, the book has shown that the creation of more opportunities for managers of small and medium sized organizations will increase the space industries innovation efforts, scope of activities, overall participants, and adopters hence the creation of more business and innovation opportunities.

Societal Implications

In the following section, we exemplify our societal implications and how our research adds value towards society.

First, many technologies that we take for granted today, initially were transferred from the space industry and any progress in any country's space sector will further add societal value through technological progression and development, an area whereby the space industry is highly significant. Second, by aiming to increase awareness among the opportunities for the Swedish space industry, we aim to foster new opportunities for employment, and for the creation of new businesses. Third, the driving force behind space investments is simply the advancement in science and technology, thus the creation of economic prosperity and the development of a more enabled technological workforce. Lastly, the creation and implementation of more business opportunities that are geared towards increased integration, and adaptation towards new space objectives would not only aim in increasing overall profitability but also aims to consider social and environmental well-being.

Moreover, the potential increase of the space industry's productivity towards more new space opportunities can increase our overall understanding of our own environment and assist in the creation of more advanced preventive measures and systems that are developed to increase humanities safety and overall well-being, as an investment in space is an investment for society and earth.

This book has explored how the Swedish space industry has evolved and the possibilities of future evolution. The change in old space mentality, lingering as far back as the cold war, has impared the possibility of significant improvements regarding small and medium sized enterprises innovation capabilities within the ecosystem.

Instead the future looks brighter and brighter, the funding organ in Sweden has identified and indeed adopted changes to improve and stimulate this very cause.

There are however a few big enterprises today that hold much power, as platform leaders that are still pushing innovation in directions that fit their strategic goals without much regard to the independence or creativity of their collaborators. The consideration of a conservative mind-set without relationships; further keeps new actors away from the market, although this is subject to change with a shift in the funding budget.

The motivational factors tell us that individuals indeed are interested in space and have realised the amount of products and services that are dependent on the space industry, more specifically downstream space. The aggregate shift in the understanding of the industry should only become more apparent moving forward, making the future of the industry more promising as a profitable market as the realisation continues to radiate.

We hope that with this book more governmental organs, big organisations and SMEs will excercise and understand how the space industry has shifted and is indeed shifting. In turn increasing coordinated efforts to include more companies in the ecosystem. Therefore, collectively furthering universal organizational capability and sustainability in humanities coordinated efforts to further our quoted dreams and ambitions for reaching the stars and beyond.

6.3 Limitations and Future research

Within this section we further express and discuss our research limitations and provide direction to areas for future research and development.

Our research was aimed to discover how the space industry is transitioning between old space and new space for the ecosystem that currently exists in Sweden, and to identify and explore the actors' stakeholder environments, and how opportunities are formed. Our sample size consisted of small to medium sized organizations and experts within the ecosystem, though we have managed to acquire data from different sides of the ecosystem, and in accordance with our research direction a limitation may be to get one of the larger organizations considered as the big players in the space industry in Sweden to provide us with their perspective.

Due to limited access from several organizations to participate in the book due to the current coronavirus pandemic, we did not have the opportunity to engage with all the entities and actors within the space ecosystem in the way we would have liked. Our research was conducted on businesses and organizations within Sweden with relative focus on the Swedish space industry, our data and results were limited only to Sweden, therefore our results will not be generalizable to other geographical regions. Furthermore, our results could possibly be transferable to other industries but due to the limitation on our sample size we feel there could be more future research conducted to cover all of the businesses and organizations operating within the space ecosystem that currently exists in Sweden.

Another area for future research would be to analyze further how platforms affect or evolve stakeholder relationships, while putting emphasis on opportunity formation between different actors within the ecosystem. We believe with a larger sample size this can be easily observed and executed. Globalization is continuously becoming the norm and changing the business landscape, opening doors to global opportunities while increasing competition.

Moreover, further research on the international collaborations that exist between different geographical regions would enhance our understanding of the global space collective and how ecosystems grow and globalize. Another critical area for future research would be to understand sustainability from the stakeholder's perspective on the global space industry, and how relationships, strategies, and exploitation of opportunities are formed and utilized towards initiatives based on environmental, and societal well-being.

Furthermore, research needs to be conducted on global private space organizations to fully understand the platform-centric relationships that exist and how they consider a platform development that embraces more actors and small businesses locally and globally.